Novel Applications of Piezoelectric and Thermoelectric Materials

Edited by Rafael Vargas-Bernal and Roberto Palma Guerrero

Published in London, United Kingdom

Novel Applications of Piezoelectric and Thermoelectric Materials
http://dx.doi.org/10.5772/intechopen.100790
Edited by Rafael Vargas-Bernal and Roberto Palma Guerrero

Contributors
Hiranmoy Samanta, Kamal Golui, Soumyadeep Mukherjee, Euripides Hatzikraniotis, George S. Polymeris, Theodora Kyratsi, Miguel Angel Olivares-Robles, Alexander Vargas Almeida, Svyatoslav Yatsyshyn, Bohdan Stadnyk, Mykola Mykyichuk, Yuriy Bobalo, Tetiana Bubela, Shinobu Aoyagi, Hiroaki, Takeda, María Sol Ruiz, Adrián César Razzitte, Tarun Garg, Lickmichand M. Goyal, Rafael Vargas-Bernal

First published in London, United Kingdom, 2024 by IntechOpen
IntechOpen is the global imprint of INTECHOPEN LIMITED, registered in England and Wales, registration number: 11086078, 5 Princes Gate Court, London, SW7 2QJ, United Kingdom

British Library Cataloguing-in-Publication Data
A catalogue record for this book is available from the British Library

Additional hard and PDF copies can be obtained from orders@intechopen.com

Novel Applications of Piezoelectric and Thermoelectric Materials
Edited by Rafael Vargas-Bernal and Roberto Palma Guerrero
p. cm.
Print ISBN 978-1-83768-396-3
Online ISBN 978-1-83768-397-0
eBook (PDF) ISBN 978-1-83768-398-7

For EU product safety concerns:
IN TECH d.o.o., Prolaz Marije Krucifikse Kozulić 3, 51000 Rijeka, Croatia,
info@intechopen.com or visit our website at intechopen.com.

Meet the editors

Rafael Vargas-Bernal earned an Electronics and Communications Engineer degree from the University of Guanajuato, Mexico, in 1995. He obtained his master's degree and a DSc in Electronics from the National Institute of Astrophysics, Optics and Electronics (INAOE), Mexico, in 1997 and 2000, respectively. Since 2000, he has worked as a full professor at the University of Guanajuato, the Instituto Tecnológico Superior de Irapuato, and the University of La Salle Bajío. He is a reviewer for numerous publishers and journals. He has published nineteen articles in indexed journals and forty-eight book chapters. He has edited two books for IntechOpen. His areas of interest are ceramic materials, nanomaterials, and gas sensors.

Roberto Palma Guerrero received a BSc, MSc, and Ph.D. in Physics from the University of Granada, Spain, in 2005, 2006, and 2012, respectively. He obtained a research grant to develop his postgraduate studies at the Department of Structural Mechanics and Hydraulic Engineering, University of Granada. From September 2010 to December 2012, he was a postdoctoral fellow at the Universitat Politécnica de Valencia in Valencia, Spain. From January 2013 to August 2019, he was an assistant lecturer at the Universitat Jaume I in Castellón, Spain. Since September 2019, he has been an associate professor at the University of Granada. Dr. Guerrero is a member of the Mecánica de Sólidos y Estructuras (MECSOL) research group.

Contents

Preface

A hot topic of research nowadays is the development and application of piezoelectric and thermoelectric materials. In addition to their direct application in energy harvesting, these materials are also being used to develop other high-added-value, multifunctional, and emerging applications. Either mechanical deformation or heat released environmentally by various processes can be utilized to generate electrical energy via the application of piezoelectric and/or thermoelectric materials.

This technical and scientific book compiles scientific research on some of the latest advances in piezoelectric and thermoelectric materials. The chapters within are organized into two sections. Section 1 includes five chapters that discuss piezoelectric materials, and Section 2 includes four chapters that discuss thermoelectric materials.

Chapter 1 introduces the characteristics and applications of two-dimensional materials, which are ideal materials for harvesting piezoelectric energy to be used in aerospace, military, automotive, and electronic clothing applications.

Chapter 2 examines the performance achieved by two-dimensional piezoelectric materials such as phosphorene, MXenes, Janus structures, heterostructured materials, and transition metal chalcogenides in energy harvesting for applications including self-powered sensors, self-powered devices, and regenerative actuators.

Chapter 3 investigates the use of time-resolved X-ray diffraction to study the amplified crystal lattice deformation and ionic displacements of resonant quartz crystals under an alternating electric field used in the implementation of piezoelectric oscillators for digital circuit applications.

Chapter 4 discusses the application of piezoelectric and magnetoelectric materials to take advantage of acoustic and mechanical vibrations, as well as magnetic fields produced by alternating current, to harvest energy.

Chapter 5 presents an equivalent circuit model of laminated magnetostrictive and/or piezoelectric composite materials to predict their behavior under dynamic electromagnetic fields for both the direct and inverse effects in the longitudinal-transverse and transverse-transverse modes.

Chapter 6 evaluates different thermal conductivity models for thermoelectric materials both phenomenologically and analytically to identify the most convenient materials to develop high-temperature applications such as thermal barrier coatings.

Chapter 7 analyzes a quantum unit as a temperature standard to develop high-precision and high-efficiency thermoelectric applications using one- and two-dimensional nanomaterials as well as nanostructured materials in energy harvesting.

Chapter 8 discusses the reduced variables methodology using computational tools in the optimal design of thermocouple legs to determine different possible sets of suitable longitudinal dimensions in the same temperature range to produce electrical power on a load.

Finally, Chapter 9 reviews the applications of thermoelectric generators in health care for wearable applications to identify the resulting operational problems and those that remain to be resolved to offer a better quality of life for patients who require energy storage devices in sensors.

This book is a valuable source of information on piezoelectric and thermoelectric materials for undergraduate and graduate students, as well as professionals, experts, and researchers interested in this cutting-edge topic and high-added value applications. The editors appreciate all the time and effort made by the authors of the research presented in this book. Likewise, we are grateful to the IntechOpen staff, especially Ms. Tea Jurcic for her valuable support throughout the editing process. Finally, the senior editor would like to thank his wife, son, parents, brothers, and sisters for their help and support.

Dr. Rafael Vargas-Bernal
Department of Communications and Electronics Engineering,
University of Guanajuato,
Salamanca, Guanajuato, México

Dr. Roberto Palma Guerrero
Department of Structural Mechanics and Hydraulic Engineering,
University of Granada,
Granada, Spain

Section 1

Piezoelectric Materials

Chapter 1

Introductory Chapter: Piezoelectric Energy Harvesting Based on Two-Dimensional Materials

Rafael Vargas-Bernal

1. Introduction

Energy harvesting (EH) is the technique of capturing and converting energy from many sources, such as mechanical load, vibrations, temperature gradients, and light, to produce relatively low levels of power in the nW-mW range [1]. The effectiveness and particular material qualities have a significant impact on the potential and performance of energy-harvesting devices. The efficiency of mechanical oscillators at tiny sizes, in addition to the materials required in the energy transduction mechanism itself, is a crucial aspect of motion energy harvesting. The crucial characteristics and uses of piezoelectric two-dimensional materials for energy-harvesting devices are covered in this chapter. Their broader applications—from the military to the aerospace industry, from cars to electronic clothing—indicate the necessity for further research on this subject.

2. Basic concepts

Energy harvesting technology is founded on the notion that devices may instantly use the energy that is present in their ambient surroundings in real time and never need to be temporarily stored [2]. This would make it possible for devices to have theoretically limitless lifespans that are only constrained by the lifespans of their constituent parts. Real-time systems must strictly adhere to predetermined reaction times to operate; hence, it must be demonstrated that this new technology is appropriate for them. The development of discrete devices that can produce electrical energy from kinetic energy (i.e., vibrations or movement), temperature gradients, or incoming light has been the main emphasis of energy harvesting up to this point [3]. The quantity and type of source energy existing in the environment have a fundamental impact on how well energy harvesting works. The limitations of the application and the specifics of the energy source must be known before building an energy-collecting solution. For each use in conventional kinetic energy harvesting, bespoke devices are frequently created, and it is uncommon to find a straightforward solution that covers a wide range of uses.

The phrase "energy harvesting" is typically used to provide electricity to small, low-power electronic components [2]. The most significant areas of use for energy harvesting are connected items, such as wireless sensors and wearable electronic

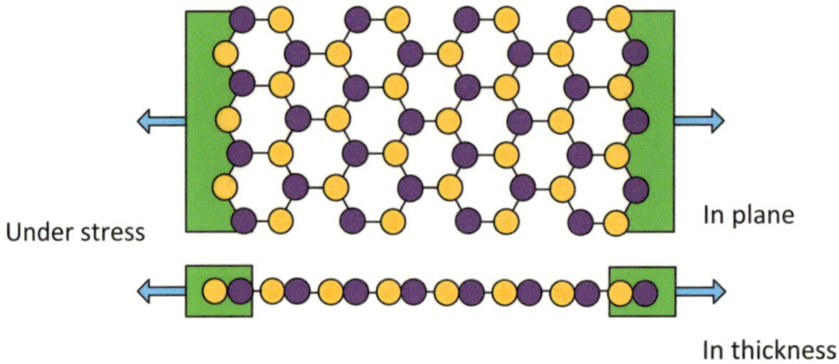

Figure 1.
The basic principle of piezoelectricity in two-dimensional materials.

devices. The use of these new technologies has noticeably led to a change in how electronic systems are designed. It poses additional difficulties for system designers, who must now try to maximize the use of ambient power to achieve energy independence in each device. There is a good chance that as time goes on, this problem will get simpler. Electronic circuitry and wireless links do use less power with time. As a result, energy harvesting technology has grown exponentially in every application area, including home automation, healthcare, the military, transportation, and others. The number of devices using ambient energy is predicted to reach 2.6 billion by 2024 [2]. Energy collecting, however, poses a unique set of difficulties, the majority of which can be attributed to the unpredictability and uncontrollability of most ambient energy sources.

The basic operating principle of a two-dimensional piezoelectric material based on hexagonal boron nitride (h-BN) is shown in **Figure 1**. The material is sensitive to mechanical stress applied by movement or vibration and is reflected in the change of interatomic distance of the chemical bonds between boron and nitrogen.

3. Novel piezoelectric materials

Piezoelectricity has historically been used in a wide range of sensing and energy-harvesting applications [4]. Nanogenerators (NGs), a new technology product that was created as a possible method for turning random mechanical energy (like human movement) into electric power, are booming in the market [5]. A prototype piezoelectric nanogenerator built of zinc oxide (ZnO) nanowires was suggested in 2006, marking an important development in the understanding and use of piezoelectricity at the nanoscale. Since then, the scientific world has become increasingly interested in piezoelectric nanomaterials, particularly piezoelectric nanowires manufactured from wurtzite-structured materials, such as ZnO and gallium nitride (GaN). The analysis and creation of NGs were then the focus of intense research employing a variety of piezoelectric materials, including ZnO, GaN, lead zirconate titanate (PTZ), and $BaTiO_3$ (BTO). Due to its advantageous optical, semiconducting, and piezoelectric properties, which make it a qualified candidate for NG fabrication, ZnO is a significant functional material [5–7]. Due to their distinctive characteristics, such as a high surface-to-volume ratio and outstanding mechanical endurance, two-dimensional (2D) ZnO nanosheets (NSs) are particularly emerging as significant nanostructures in the construction process of high-performance NGs. Recently, a study was conducted to compare the effects of employing curved (CNSs) and straight nanosheets (SNSs). It was found that the piezopotentials produced by the two nanosheets in the event of vertical compression differed slightly in magnitude and distribution, however, in the case of lateral bending, the piezopotential produced by CNSs was consistently significantly bigger than that produced by SNSs [7]. On applying 4 kg of force for more than 4000 cycles, the optimized NGs generated voltages, current densities, and power densities of up to 0.9 V, 16.5 A/cm^2, and 600 nW/cm^2, respectively [6].

In recent years, two-dimensional (2D) materials, including hexagonal boron nitride (*h*-BN), and many transition-metal dichalcogenides (TMDCs), transition-metal dioxides (TMDOs), II-VI semiconductors, III-V semiconductors, III-monochalcogenides and IV-monochalcogenides, and others, have also been found to exhibit piezoelectricity and the piezotronic effect [8, 9]. Two-dimensional materials are discovered to have numerous distinct piezoelectric capabilities in comparison to traditional piezoelectric nanowires. For instance, some 2D materials, including monolayer group IV monochalcogenides, have enormous piezoelectric coefficients that are two orders of magnitude greater than those of traditional ZnO and GaN nanowires. Simulations and experiments also demonstrate that by altering the number of layers of 2D materials, it is possible to effectively change their piezoelectric performance. Since their development, 2D materials have attracted a lot of attention for piezotronic applications because of their special benefits in terms of their piezoelectricity and their capacity to endure extreme strain.

For instance, many 2D materials, such as the thin-layered MoS_2, only exhibit piezoelectric effects in their odd-numbered layers, whereas their even layers only exhibit a modest response [8]. Additionally, the device system is limited in its ability to access out-of-plane activities since the piezoelectricity in many 2D materials is mostly detected within the in-plane piezoelectric polarization. Non-centrosymmetry, functionalization, and doping can be induced in the resulting crystal lattice of the two-dimensional material with a higher degree of anisotropy by reducing the thickness, which also breaks the crystal's symmetry. Other types of 2D layered materials that may be of interest include post-transition metal dichalcogenides (PTMDs), such as SnS_2, which have been confirmed theoretically but have yet to be fully explored experimentally [8].

New energy-harvesting technologies are based on 2D materials and utilize a variety of mechanisms, including photovoltaic, thermoelectric, piezoelectric, triboelectric, and hydrovoltaic devices. Additionally, advances in the harvesting of osmotic pressure and wireless energy have been made [10].

In contrast to graphene, which has identical atoms and a center of symmetry, monolayers of *h*-BN, MoS_2, WSe_2, and TMDCs exhibit significant piezoelectricity because their inversion symmetry is broken [11]. To create cantilever energy harvesters, layered 2D nanosheets of *h*-BN and MX_2 (M = Mo or W; X = S, Se, or Te) have been coated on a silicon substrate [12]. Non-centrosymmetric two-dimensional (2D) layered materials have a lot of potential for nanoscale electromechanical systems and electronic devices [13]. Due to their mechanical flexibility and piezoelectricity, some 2D materials have the potential to be employed in nanodevices that can generate wearable power. By adding strain to the structure, the bandgap can be tuned. The bandgap of trilayer MoS_2 has been found to grow by 300 meV for every 1% biaxial compressive strain. A vertical electric field application has also been proposed to potentially zero out the bandgap in 2D TMDCs, converting the structure from semiconducting to metallic [14].

The Janus monolayers lack the reflection symmetry regarding the core metal atoms, which permits out-of-plane electric polarizations, in contrast to ideal 2D

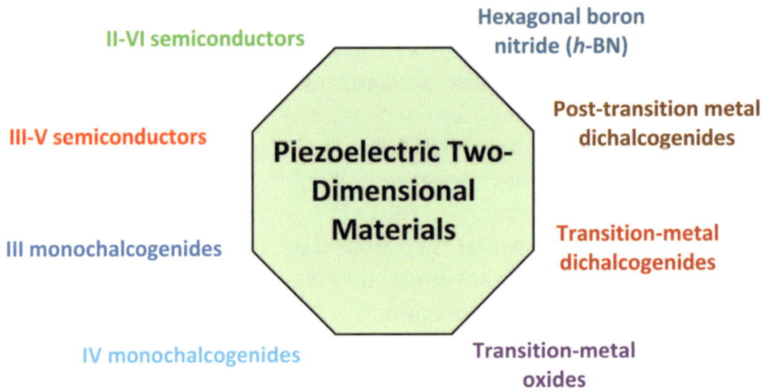

Figure 2.
Main two-dimensional piezoelectric materials.

materials. Future atomically thin piezoelectric applications in the fields of transducers, sensors, and energy harvesting devices should favor 2D tin dichalcogenide-derived Janus monolayers [15].

A compilation of two-dimensional materials with piezoelectric properties is illustrated in **Figure 2**.

4. Applications

Significant efforts have been made to produce sustainable, mobile, and distributed power sources for the energy of a new era due to the rapid development of portable devices, wearable electronics, and the Internet of Things. The rising energy needs, particularly for portable and wearable devices, may be met by scavenging the otherwise wasted energy from the ambient environment into electrical power [10]. Future wireless nanosystems without power supply, such as environmental monitoring, implanted medical sensors, and personal electronics, will depend on energy-harvesting cells based on 2D piezoelectric materials [13].

Sn-based chalcogenide Janus structures have a potential future for atomically thin applications in mechanical sensors, energy sensors, etc., thanks to their strong piezoelectric characteristics [15]. Despite the advancements made to date, 2D materials energy-harvesting devices still face numerous obstacles and opportunities in the creation of self-powered electronics and wearable technology before widespread use is even conceivable.

It has been demonstrated that flexible PZT piezoelectric energy harvesters based on mica can capture energy from muscle movements. A workable approach to self-powered bio-devices is to scavenge mechanical energy from the motions of people or internal organs [16]. It demonstrated excellent mechanical and electrical durability by being continually bent and unbent at a high strain 40,000 times without obviously degrading output performance.

Numerous materials are theoretically anticipated to be piezoelectric in their 2D forms due to the 2D confinement and spontaneous breaking of inversion symmetry [17]. A new platform for the investigation of novel physics at the atomic scale as well as prospective device applications is created by the combination of piezoelectricity and other fascinating features in 2D materials. Although the basic idea behind

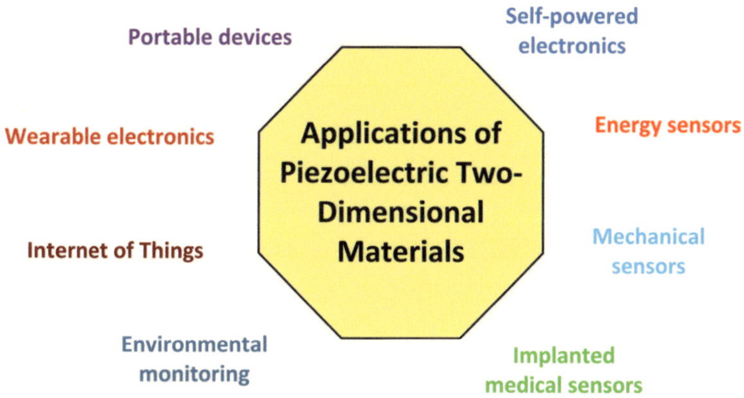

Figure 3.
Main applications of piezoelectric two-dimensional materials.

piezotronic and piezophototronic phenomena is widely understood, further research is still needed to better understand how piezoelectricity interacts with semiconductor characteristics.

A compilation of the applications of two-dimensional materials with piezoelectric properties is depicted in **Figure 3**.

5. Conclusions

Piezoelectricity is a characteristic of macroscopic strain-induced electric polarization that couples mechanical and electrical characteristics to enable effective mechanical-to-electrical energy transfer. The piezoelectric effect, which has benefits like high efficiency, lightweight, and tiny scale, is the most alluring method for mechanical-to-electrical energy conversion. Numerous 2D materials have been anticipated to have intrinsic piezoelectricity, however, due to the enormous challenges in material synthesis, most of them have not yet been empirically examined. Modern semiconductor processes, as well as classic electronic technologies, are simple to integrate with 2D piezoelectric materials. The following aspects of 2D piezoelectric materials need to be studied to achieve their commercial success: the impact of strain on an electronic structure on piezoelectric charges, how carrier concentrations can improve the piezoelectric output, and the search for more chemically stable piezoelectric materials under ambient conditions.

Acknowledgements

The author appreciates the support of the University of Guanajuato to develop this research.

Thanks

The author wants to thank his wife and son for their support and time to edit this book. The author appreciates the support of Tea Jurcic working for IntechOpen as an author service manager.

Author details

Rafael Vargas-Bernal
University of Guanajuato, Salamanca, Guanajuato, México

*Address all correspondence to: rvargasbernal@gmail.com

References

[1] Soin N. Magnetic nanoparticles – Piezoelectric polymer nanocomposites for energy harvesting. In: El-Gendy AA, Barandiarán JM, Hadimani RL, editors. Magnetic Nanostructured Materials: From Lab to Fab. Cambridge, MA: Elsevier; 2018. pp. 295-322. DOI: 10.1016/B978-0-12-813904-2.00010-3

[2] Chetto M, Queudet A. Harnessing ambient energy for embedded systems. In: Energy Autonomy of Real-Time Systems. London, UK: ISTE Press-Elsevier; 2016. pp. 59-83. DOI: 10.1016/B978-1-78548-125-3.50003-8

[3] Beeby SP, Cao Z, Almussallam A. Kinetic, thermoelectric and solar energy harvesting technologies for smart textiles. In: Kirstein T, editor. Multidisciplinary Know-How for Smart-Textiles Developers. Cambridge, UK: Woodhead Publishing; 2013. pp. 306-328. DOI: 10.1533/9780857093530.2.306

[4] Zhang J. On the Piezopotential properties of two-dimensional materials. Nano Energy. 2019;**58**:568-578. DOI: 10.1016/j.nanoen.2019.01.086

[5] Wang Q, Yang D, Qiu Y, Zhang X, Song W, Hu L. Two-dimensional ZnO nanosheets grown on flexible ITO-PET substrate for self-powered energy-harvesting nanodevices. Applied Physics Letters. 2018;**112**(6):063906. DOI: 10.1063/1.5012950

[6] Lee Y, Kim S, Kim D, Lee C, Park H, Lee J-H. Direct-current flexible piezoelectric nanogenerators based on two-dimensional ZnO nanosheet. Applied Surface Science. 2020;**509**:145328. DOI: 10.1016/j.apsusc.2020.145328

[7] Yang A, Qiu Y, Yang D, Lin K, Guo S. Piezoelectric property comparison of two-dimensional ZnO nanostructures for energy harvesting devices. RSC Advances. 2021;**11**(6):3363-3370. DOI: 10.1039/d0ra10371c

[8] Wang Y, Vu L-M, Lu T, Xu C, Liu Y, Ou Z, et al. Piezoelectric responses of mechanically exfoliated two-dimensional SnS_2 nanosheets. ACS Applied Materials & Interfaces. 2020;**12**(46):51662-51668. DOI: 10.1021/acsami.0c16039

[9] Hinchet R, Khan U, Falconi U, Kim S-W. Piezoelectric properties in two-dimensional materials: Simulations and experiments. Materials Today. 2020;**21**(6):611-630. DOI: 10.1016/j.mattod.2018.01.031

[10] Fan FR, Wu W. Emerging devices based on two-dimensional monolayer materials for energy harvesting. Research. 2019;**2019**:7367828. DOI: 10.34133/7367828

[11] Lee G-J, Lee M-K, Park J-J, Hyeon DY, Jeong CK, Park K-I. Piezoelectric energy harvesting from two-dimensional boron nitride nanoflakes. ACS Applied Materials & Interfaces. 2019;**11**(41):37920-37926. DOI: 10.1021/acsami.9b12187

[12] Nan Y, Tan D, Shao J, Willatzen M, Wang ZL. 2D materials as effective cantilever piezoelectric Nano energy harvesters. ACS Energy Letters. 2021;**6**(6):2313-2319. DOI: 10.1021/acsenergylett.1c00901

[13] Cui C, Xue F, Hu W-J, Li L-J. Two-dimensional materials with piezoelectric and ferroelectric functionalities. NPJ 2D Materials and Applications. 2014;**2**:18. DOI: 10.1038/s41699-018-0063-5

[14] Tahir MB, Fatima U. Recent trends and emerging challenges in

two-dimensional materials for energy harvesting and storage applications. Energy Storage. 2021;**4**(1):e244. DOI: 10.1002/est2.244

[15] Zhang X, Cui Y, Sun L, Du J, Huang Y. Stabilities, and electronic and piezoelectric properties of two-dimensional tin dichalcogenide derived Janus monolayers. Journal of Materials Chemistry C. 2019;**7**(42):13203-13210. DOI: 10.1039/c9tc04461b

[16] Wang D, Yuan G, Hao G, Wang Y. All-inorganic flexible piezoelectric energy harvester enabled by two-dimensional Mica. Nano Energy. 2018;**43**:351-358. DOI: 10.1016/j.nanoen.2017.11.037

[17] Lin P, Pan C, Wang ZL. Two-dimensional nanomaterials for novel piezotronics and piezophototronics. Materials Today Nano. 2018;**4**:17-31. DOI: 10.1016/j.mtnano.2018.11.006

Chapter 2

Advances in Piezoelectric Two-Dimensional Materials for Energy Harvesting

Rafael Vargas-Bernal

Abstract

The design of piezoelectric energy harvesting systems can be exploited for the development of self-powered sensors, human-powered devices, and regenerative actuators, as well as the development of self-sustained systems with renewable resources. With the introduction of two-dimensional materials, it is possible to implement piezoelectric nanostructures to exploit environmental energies, taking advantage of their flexible mechanical structures. This chapter aims to study the relevant contribution that piezoelectric two-dimensional materials have in energy harvesting. Among the two-dimensional piezoelectric materials analyzed are phosphorene, MXenes, Janus structures, heterostructured materials, and transition metal dichalcogenides (TMDs). These materials are studied through their performance from a piezoelectric point of view. The performance achieved by two-dimensional piezoelectric materials is comparable to or even better than that achieved by bulk piezoelectric materials. Despite the advances achieved so far, many more materials, as well as structures for the implementation of energy harvesting devices or systems, will be proposed in this century, so this research topic will continue to be interesting for research groups around the world.

Keywords: two-dimensional materials, piezoelectric materials, transition metal dichalcogenides (TMDs), heterostructures, energy harvesting, morphotropic phase boundary (MPB), Janus structures, MXenes, phosphorene

1. Introduction

The direct piezoelectric effect was discovered in 1880 by Pierre and Jacques Curie in quartz crystal (silicon dioxide, SiO_2) [1]. In the direct piezoelectric effect, it is possible to generate an electrical charge proportional to the applied mechanical effort. In the reverse piezoelectric effect, a proportional geometric deformation is achieved by an applied voltage [1]. It was not until 1947 that Shepard Roberts that the first polycrystalline piezoelectric ceramic based on barium titanate ($BaTiO_3$) was discovered to exhibit 100 times more piezoelectricity than quartz. In the 1950s, it was discovered that other oxides such as lead titanate ($PbTiO_3$) and lead zirconate ($PbZrO_3$) have twice the piezoelectric properties of barium titanate. Each material has a Curie temperature above which piezoelectricity disappears. In addition, for each polycrystalline piezoelectric material, there is a cation ratio that must be optimized to reach the morphotropic phase boundary (MPB), which produces the presence of rhombohedral and tetragonal phases that allow adjusting the piezoelectric properties. In a piezoelectric material, through the applied mechanical stress, it is possible to achieve a total separation of positive and negative charges thanks to the non-centrosymmetric structure of the piezoelectric material. Due to the harmful influence of lead found in piezoelectric materials on the safety of workers who process these materials, as well as the damage to soil and water, researchers are investigating the possibility of developing lead-free piezoelectric materials. Piezoelectric materials are mainly applied to energy harvesting and sensing [2–36]. Without a doubt, one of the contributions that motivated energy harvesting was the use of zinc oxide (ZnO) nanowires as a piezoelectric nanogenerator using nanomaterials [26]. Through the use of two-dimensional materials, it is possible to exploit both semiconductor properties and piezoelectric properties for the tuning and transport of charge carriers [2–36].

The electromechanical interaction in these materials is related to a linear constant between the displacement or electric field and the stress or strain achieved mechanically called piezoelectric coefficient and expressed as C/N or m/V [1]. The piezoelectric coefficients of materials are not uniform along the crystalline directions, that is, are anisotropic, so it is necessary to consider the crystalline direction that will be used in the design of devices that take advantage of piezoelectric properties. Piezoelectric materials enable the development of applications such as energy harvesting in living environments by creating self-powered sources to power electronic devices at the nanoscale. To assess the piezoelectric performance of a material, four main piezoelectric coefficients can be distinguished: the piezoelectric charge coefficient or piezoelectric strain coefficient (d), the piezoelectric voltage coefficient or voltage output constant (g), the piezoelectric stress coefficient (e), and the piezoelectric stiffness coefficient (h). Of the 32 possible crystalline solid groups, there are 20 piezoelectric groups and 10 ferroelectric groups. All ferroelectric materials have pyroelectric behavior and all of them are piezoelectric; by contrast, piezoelectric materials are not necessarily pyroelectric or ferroelectric. The most frequently used piezoelectric coefficient is d and it relates the open circuit charge density to the applied mechanical stress and regularly uses two subscripts. The first subscript states the direction in which the electrical polarization is induced, and the second subscript specifies the direction in which the stress is applied. This relates the electromechanical coupling coefficient k, to the relative dielectric constant k^{T} at a constant stress, and the elastic compliance at a constant electric field s^{E}, as well as the electric permittivity of the vacuum ε_0, and this is expressed mathematically $asd = k\sqrt{\varepsilon_0 k^T s^E}$. The two most important d constants in the case of nanomaterials are d_{31} (induced strain in direction 1 per unit electric field applied in direction 3) and d_{11} (induced strain in direction 1 per unit electric field applied in direction 1). The piezoelectric stress coefficient (e) can also be defined for the subscripts 11 and 31 of the material for the reverse piezoelectric behavior. For two-dimensional materials, subscripts 11 are considered the in-plane piezoelectric response, while subscripts 31 and 33 are used for the out-of-plane piezoelectric response.

The rest of this chapter has been organized as follows: In Section 2, a summary of the basic concepts associated with piezoelectricity and two-dimensional materials is presented. Next, novel piezoelectric materials are examined using comparative tables and graphs of their in-plane and out-of-plane performance in Section 3. In Section 4, applications of two-dimensional materials in energy harvesting and related topics are described. Future research directions related to these materials are discussed in Section 5. Finally, the conclusions of this chapter are presented.

2. Basic concepts

A two-dimensional material is a crystalline solid with at least one of its dimensions on the nanoscale. In 2004, graphene was the first two-dimensional material to be discovered, and this is a carbon-based material with a thickness of one atom [2]. Two-dimensional layered materials can be classified as homoatomic when they only contain one chemical element in their structure, such as graphene, phosphorene, antimonene, and heteroatomic when they contain more than 1 chemical element in their structure as in the case of hexagonal boron nitride (*h*-BN), α-phase indium sulfide (α-In_2S_3), gallium arsenide (GaAs), etc. Two-dimensional layered materials with homostructure are naturally centrosymmetric because they do not show out-of-plane piezoelectricity and have in-plane piezoelectricity; however, this quality depends on the degree of ionization of their covalent bonds. The advantage of two-dimensional layered materials with heterostructure exhibits naturally in-plane and out-of-plane piezoelectricity. Two-dimensional materials are arranged in layers that are stacked and held together through weak van der Waals interactions. Since the discovery of graphene, a large set of two-dimensional materials based on the chemical elements of groups IVA, IIIA-VA, IIB-VIA, and other chalcogenide metals of group B have been discovered and proposed. These materials exhibit unique chemical and physical properties that sometimes differ from those of their bulky counterparts [3, 6–24]. Two-dimensional layered materials are interesting options for implementing energy conversion, harvesting, and storage applications due to their high surface area, processing, and assembly versatility, as well as multiple novel surface chemistries. In these materials, the d_{33} and g_{33} piezoelectric coefficients corresponding to out-of-plane piezoelectricity have zero or none, while the d_{31} and g_{31} piezoelectric coefficients corresponding to in-plane to out-of-plane piezoelectricity have a modest value. Two-dimensional materials to be exploited piezoelectrically must have a large surface-to-volume ratio, extraordinary electrical properties, and extremely low noise level, which also lead them to be used in sensing applications. When comparing sensors based on microelectromechanical systems (MEMS) with sensors based on two-dimensional materials, the latter are much lighter and smaller, have lower power consumption, have more flexibility, and show superior sensitivity. In addition to reducing the thickness of the two-dimensional material, other strategies that can be used to provide these same non-centrosymmetry properties are functionalization and doping of the crystal lattice to favor its anisotropy and thereby achieve better piezoelectric properties [11]. Other approaches that are being used to increase the piezoelectric properties are the generation of defects in the layers or the implementation of heterostructured materials based on two-dimensional materials. Because all the atoms in two-dimensional materials are exposed on their surface, it is possible to tune their physical properties. Within these properties, piezoelectric properties can be used to maximize energy harvest using micro- and nanoelectromechanical systems. Applications of nanoscale piezoelectric materials are transducers, actuators, and energy harvesters for fields such as piezotronics and nanorobotics.

For a material to present piezoelectric properties, the unit cell must not have a symmetrical center so that an anisotropic dipole moment occurs, which is called dielectric polarization [3]. This phenomenon is observed in crystalline and

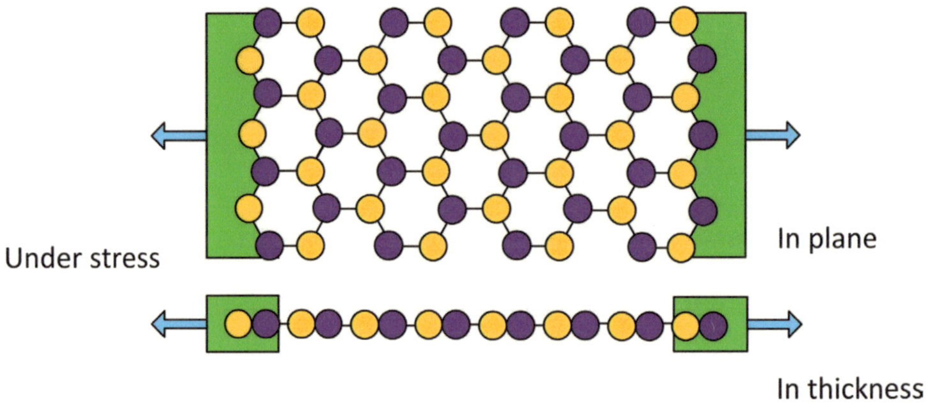

Figure 1.
The basic principle of piezoelectricity in two-dimensional materials.

semicrystalline dielectric materials when an electric field is applied. The polarization orients the cations and anions found in the material either partially or completely in the direction of the field, the higher the orientation the better the piezoelectric coefficient of the material, especially when this response is achieved under a continuously oscillating field. Traditional applications of piezoelectric materials are in microbalances, high-resolution mechanical actuators, and quartz oscillators [28].

Wearable nanogenerators to generate power can be implemented using mechanically flexible materials that present piezoelectric properties such as two-dimensional materials [2]. The piezoelectric properties of transition metal dichalcogenides (TMDs) appear only in monolayers and disappear in bilayers. So far, transition metal dichalcogenides (TMDs) with odd layer numbers have piezoelectric properties due to the absence of inversion symmetry. When the two-dimensional material is mechanically bent at both ends, the nanosheet is expanded, which causes the polarized charges to deliver a flow of electrons toward an external charge, as illustrated in **Figure 1**. When the two-dimensional material is mechanically released, the electron flow stops. The periodic stretching and releasing of the two-dimensional material can produce an alternating piezoelectric output. This output will be capable of generating a voltage that can be exploited to harvest energy from a source that can generate an oscillating mechanical stress on the two-dimensional material.

3. Novel piezoelectric materials

In Ref. [4] the authors using a data mining algorithm of more than 50,000 inorganic crystals identified that there are 1173 two-dimensional layered materials. Three hundred twenty-five of these materials have piezoelectric monolayers [4]. They also found that there are 98 loosely bound Van der Waals heterostructures. According to Refs. [3, 6], different types of two-dimensional materials with piezoelectric properties can be distinguished. The first type is made up of dichalcogenides based on conventional and Janus-type transition metals. The second type is based on compounds based on the elements of groups IIA and VIA. The third type is based on compounds based on the elements of groups IIIA and VA. The fourth type is made up of compounds based on conventional and Janus-type group IIIA-VIA elements. The fifth type is made up of compounds based on the elements of the IVA and VIA groups.

Finally, the sixth type is constituted by the compounds based on two elements of the VA group. To produce a non-centrosymmetric structure in materials, it is necessary to use methodologies to modify the interfacial interaction between the ions or layers of the two-dimensional material, produce atomic adsorption on the surface, and/or introduce different defects that can modify the piezoelectric properties [3].

Transition metal dichalcogenides have the MX_2 structure based on two different chemical elements, a transition metal M and a chalcogen X atom [3]. Transition metal dichalcogenides have an ultra-thin thickness, tunable bandgap, and unique mechanical, optical, and electronic properties [5]. Multiple derivatives of 2D materials can be obtained through alloying and van der Waals heterojunctions. The symmetry of two-dimensional materials can be modified through the application of stress or electric field, growth on substrates, or the formation of heterojunctions. Symmetry breaking can be achieved using the Janus structure by modifying chalcogenides based on chemical elements of groups VIA and/or IIIA. The Janus structure is obtained by changing the monolayer MX_2 by the MXY structure where X, Y = S, Se, and Te, and X is different from Y. This causes the electrical charge distributions between the M-X and M-Y layers to differ because the atomic radii and electronegativities are different for elements X and Y. Examples of these materials are the stable monolayers MoSSe, WSSe, WSeTe, and WSTe and the unstable monolayers MoSeTe and MoSTe.

Atomic layer substitution or Janus substitution can also be used on group IV monochalcogenides to break their symmetry to enrich optical and electrical properties [6]. Group IV two-dimensional monochalcogenides include the following materials: germanium sulfide (GeS_2), germanium selenide ($GeSe_2$), germanium telluride ($GeTe_2$), tin sulfide (SnS_2), tin selenide ($SnSe_2$), and tin telluride ($SnTe_2$). Due to the geometry of the parent materials, Janus substitution can be performed in three different ways: (1) A chalcogen (S, Se, or Te) is replaced to produce a ternary material, (2) a crystallogen (Ge) is replaced or Sn to produce a ternary material, and (3) both a chalcogen and a crystallogen are replaced to produce a quaternary material. With Janus substitution, 15 two-dimensional materials of group IV monochalcogenides can be produced: Ge_2SSe, Ge_2STe, Ge_2SeTe, Sn_2SSe, Sn_2STe, Sn_2SeTe, $GeSnS_2$, $GeSnSe_2$, $GeSnTe_2$, GeS/SnSe, GeS/SnTe, GeSe/SnS, GeSe/SnTe, GeTe/SnS, and GeTe/SnSe. These materials present mechanical stability, dynamic stability, and energetic stability. Only GeS/SnTe provides dynamic instability. These materials exhibit high piezoelectric coefficients, direct-to-indirect band transitions, as well as high figures of merit for thermoelectric effects. The symmetry breaking produced by Janus substitution leads to high vertical piezoelectric coefficients to increase the efficiency of energy harvesting and other applications that will be described in the chapter.

In the case of transition metal dichalcogenides, the d_{11} piezoelectric coefficient ranges from 2.12 to 13.54 pm/V [3]. In the case of Janus-type transition metal dichalcogenides, the range of the in-plane d_{11} piezoelectric coefficient is from 2.26 to 5.30 mp/V, and for the out-of-plane d_{31} piezoelectric coefficient is from 0.007 to 0.30 pm/V. The d_{11} piezoelectric coefficient for the IIA-VIA compounds ranges from −1.16 to 26.7 pm/V, and this value increases as the atomic number of the group IIA ion decreases and it decreases as the atomic number of the group VIA ion increases. In the case of IIIA-VA compounds with in-plane structure, the d_{11} piezoelectric coefficient ranges from 0.09 to 5.5 pm/V. For these same materials with a non-planar structure, the d_{11} piezoelectric coefficient ranges from 0.02 to 1.50 pm/V and the d_{31} piezoelectric coefficient ranges from 0.02 to 0.57 pm/V. The values of both coefficients increase with decreasing the atomic number of group VA atoms or increasing the atomic number of the group IIIA atom except for aluminum, gallium, and indium nitrides.

For the IIIA-VIA compounds, the values of the d_{11} piezoelectric coefficient are in the range of 1.12 to 1.98 pm/V. The values of the d_{11} and d_{31} piezoelectric coefficients for the IIIA-VIA Janus-type compounds range from 1.91 to 8.47 pm/V and from 0.07 to 0.46 pm/V, respectively. In the case of IVA-VIA compounds, the values of the d_{11} piezoelectric coefficient range from 75.43 to 250.28 pm/V for a zigzag orientation. For these materials but with armchair orientation, the value of the d_{11} piezoelectric coefficient is in the range of 20.7 to 921.56 pm/V. In the case of a hexagonal orientation, the d_{11} piezoelectric coefficient ranges from −5.65 to −4.63 pm/V. Furthermore, the e_{14} piezoelectric coefficient ranges from 345 to 381 pC/N. Finally, in the VA-VA compounds the alpha (α) phase and the beta (β) phase occur, and the d_{11} piezoelectric coefficients range from 6.94 to 243.45 pm/V and from 0.67 to 4.83 pm/V, respectively. The values of the piezoelectric coefficients for both phases increase as the atomic number of the VA ion increases.

Two-dimensional piezoelectric nanosheets that have been proposed for energy sensing and harvesting applications include hexagonal boron nitride (h-BN), molybdenum sulfide (MoS_2), and tungsten diselenide (WSe_2) by exploiting their excellent mechanical flexibility and excellent piezoelectric response [10]. The use of two-dimensional materials with an odd number of layers allows the inversion of symmetry to be broken, which leads to strong piezoelectricity. For very specific cases, it has been shown that an even number of layers can reach considerable piezoelectric coefficients [11]. Even the armchair or zigzag crystal orientation of the layers of the two-dimensional material used in the piezoelectric material influences the value of the achieved piezoelectric coefficient [12]. In addition to traditional two-dimensional materials, novel emerging materials based on post-transition metals such as tin disulfide (SnS_2) are being proposed as piezoelectric materials. These materials present an out-of-plane piezoelectric response of 2 ± 0.22 pm/V for a thickness of 4 nm, whose value is higher than that obtained for materials such as lithium niobate ($LiNbO_3$) whose average value is around 1 pm/V.

The authors in Ref. [7] have estimated through *ab initio* simulation using density-functional theory the piezoelectric coefficients of 37 two-dimensional materials among which are transition metal dichalcogenides, oxides, and III-V compounds. The values obtained in Ref. [7] have been plotted through **Figures 2–5**, to visualize the wide variety of possible piezoelectric two-dimensional materials. **Figure 2** shows the values reached for the coefficients d_{11} and e_{11} in the plane for two-dimensional transition metal dichalcogenides (TMDs) with a $2H$ phase. The d_{11} coefficient for the dichalcogenides reaches values ranging from 2.12 pm/V for WS_2 to 13.45 pm/V for $CdTe_2$. In the case of the e_{11} coefficient for the same materials, it ranges from 184

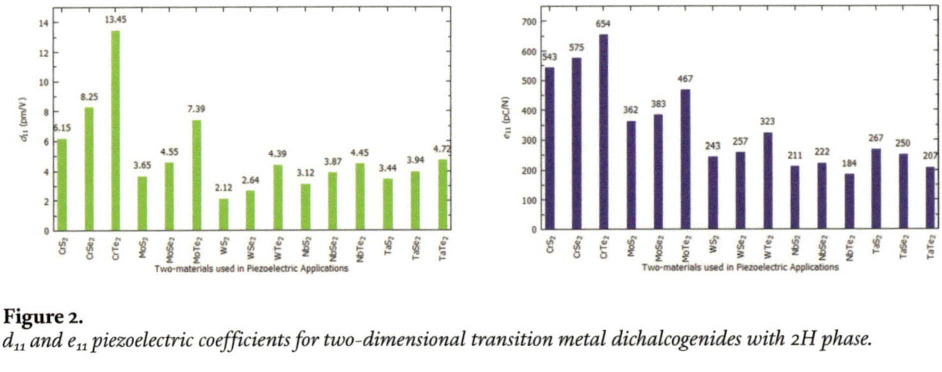

Figure 2.
d_{11} and e_{11} piezoelectric coefficients for two-dimensional transition metal dichalcogenides with 2H phase.

DOI: http://dx.doi.org/10.5772/intechopen.113754

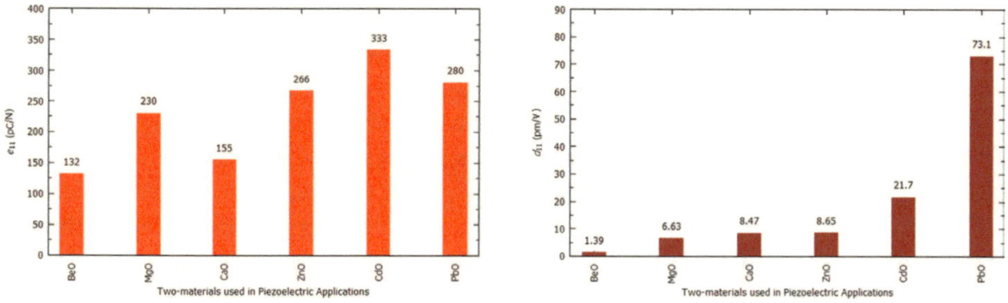

Figure 3.
d_{11} and e_{11} piezoelectric coefficients for two-dimensional II-VI oxides.

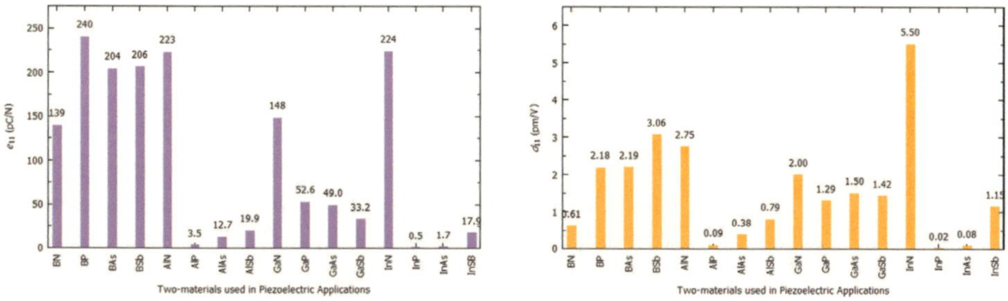

Figure 4.
d_{11} and e_{11} piezoelectric coefficients for two-dimensional III-V compounds.

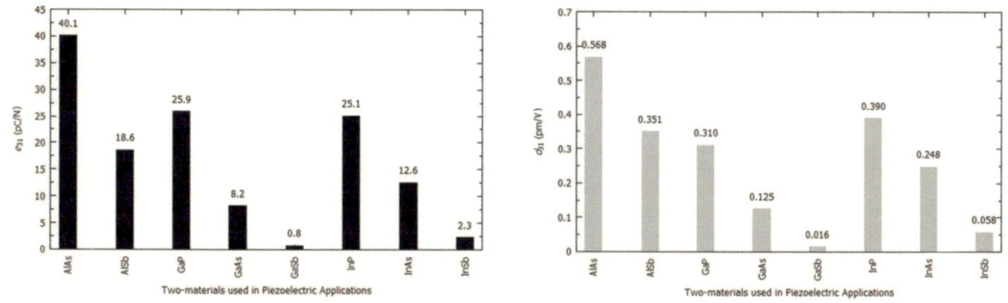

Figure 5.
d_{31} and e_{31} piezoelectric coefficients for two-dimensional III-V compounds.

pC/N for $NbTe_2$ to 654 pC/N for $CrTe_2$. The values of the piezoelectric coefficients d_{11} are very similar to those achieved by bulk piezoelectric materials such as quartz-alpha (α-SiO_2) (2.3 pm/V), gallium nitride (GaN) (3.1 pm/V), and aluminum nitride (AlN) (5.1 pm/V). Most volumetric lead zirconates (PZT) have piezoelectric coefficients d_{11} on the order of 360 pm/V. Volumetric molybdenum disulfide (MoS_2) has a coefficient e_{11} of 290 pC/N.

Figure 3 depicts the values reached for in the plane coefficients d_{11} and e_{11} in the plane for different two-dimensional II-VI oxides with primitive (p) structures. The d_{11} coefficient for the dichalcogenides reaches values ranging from 1.39 pm/V for BeO to 73.1 pm/V for PbO. In the case of the e_{11} coefficient for the same materials, it ranges from 132 pC/N for BeO to 333 pC/N for CdO.

Figure 4 illustrates the values reached for in the plane coefficients d_{11} and e_{11} in the plane for some two-dimensional III-V compounds with primitive (p) or blende

(b) structures. The d_{11} coefficient for dichalcogenides reaches values ranging from 0.02 pm/V for InP to 5.50 pm/V for InN. In the case of the e_{11} coefficient for the same materials, it ranges from 0.5 pC/N for InP to 240 pC/N for BP.

Figure 5 shows the values reached for the out-of-plane coefficients d_{31} and e_{31} in the plane for different two-dimensional III-V compounds with hexagonal structures. The d_{11} coefficient for dichalcogenides reaches values ranging from 0.016 pm/V for GaSb to 0.568 pm/V for AlAs. In the case of the e_{11} coefficient for the same materials, it ranges from 0.8 pC/N for GaSb to 40.1 pC/N for AlAs.

The authors in Ref. [8] have predicted through first-principles calculations some piezoelectric coefficients of III-V compounds for gallium arsenide (GaAs), gallium phosphide (GaP), gallium antimonide (GaSb), indium arsenide (InAs), indium phosphide (InP), and indium antimonide (InSb). The simulations consider in their determination both clamped ions and relaxed ions [8]. In **Tables 1–4**, to study the behavior of piezoelectric two-dimensional III-V compounds, comparisons are made between the values obtained by Refs. [7, 8] for the piezoelectric coefficients e_{11}, e_{31}, d_{11}, and d_{31}, respectively. The simulated values illustrate trends related to the assumptions about the theoretical principle and the mechanical physical state in which the ions are simulated to operate, which do not necessarily illustrate the actual experimentally obtained piezoelectric coefficient but serve to confirm the relevance of using two-dimensional materials in energy harvesting compared to other conventional materials already used for years.

The authors of Ref. [9] used density functional perturbation theory (DFPT) and first-principles calculations to predict the piezoelectric coefficients of

Nanomaterial	e_{11} [7]	e_{11} (clamped-ion) [8]	e_{11} (relaxed-ion) [8]
GaP	25.9	307	44
GaAs	49.0	289	33
GaSb	33.2	254	25
InP	0.5	290	40
InAs	1.7	281	32
InSb	17.9	252	24

Table 1.
Compilation of e_{11} piezoelectric coefficients for some two-dimensional III-V compounds predicted by different simulation techniques.

Nanomaterial	e_{31} [7]	e_{31} (clamped-ion) [8]	e_{31} (relaxed-ion) [8]
GaP	52.6	59	450
GaAs	8.20	90	280
GaSb	0.80	69	190
InP	25.1	100	480
InAs	12.6	130	310
InSb	2.3	84	170

Table 2.
Compilation of e_{31} piezoelectric coefficients for some two-dimensional III-V compounds predicted by different simulation techniques.

Nanomaterial	d_{11} [7]	d_{11} (clamped-ion) [8]	d_{11} (relaxed-ion) [8]
GaP	1.29	5.10	0.96
GaAs	1.50	6.06	0.93
GaSb	1.42	6.91	0.90
InP	0.02	6.33	1.45
InAs	0.08	7.48	1.20
InSb	1.15	8.36	1.16

Table 3.
Compilation of d_{11} piezoelectric coefficients for some two-dimensional III-V compounds predicted by different simulation techniques.

Nanomaterial	d_{31} [7]	d_{31} (clamped-ion) [8]	d_{31} (relaxed-ion) [8]
GaP	0.310	0.062	0.51
GaAs	0.125	0.12	0.40
GaSb	0.016	0.12	0.33
InP	0.390	0.13	0.74
InAs	0.248	0.20	0.51
InSb	0.058	0.17	0.35

Table 4.
Compilation of d_{31} piezoelectric coefficients for some two-dimensional III-V compounds predicted by different simulation techniques.

Nanomaterial	d_{11} [7]	d_{11} (clamped-ion) [9]	d_{11} (relaxed-ion) [9]	d_{11} (DFPT) [9]
BeO	1.39	1.25	4.62	1.43
MgO	6.63	18.2	5.00	7.69
CaO	8.47	9.63	10.2	10.2
SrO	—	12.6	13.3	7.40
BaO	—	27.4	24.5	1.07
ZnO	8.65	5.90	9.29	8.57
CdO	21.7	8.57	8.21	23.4

Table 5.
Compilation of d_{11} piezoelectric coefficients for some two-dimensional IIA/IIB-VI oxides predicted by different simulation techniques.

two-dimensional oxides of IIA/IIB groups. **Tables 5** and **6** present compilations of the values obtained for the d_{11} and e_{11} piezoelectric coefficients. The simulated values vary depending on the simulation conditions and these are very similar considering the second and last columns, but not for all materials. Through the tables and figures in this section, it can be broken down that the values obtained in the simulations must be verified because similar values, or with significant differences, were obtained.

The values are attractive from the point of view that these materials allow the implementation of applications where out-of-plane piezoelectric coefficients can be

Nanomaterial	e_{11} [7]	e_{11} (clamped-ion) [9]	e_{11} (relaxed-ion) [9]	e_{11} (DFPT) [9]
BeO	132	123	385	146
MgO	230	557	152	236
CaO	155	152	167	156
SrO	—	156	167	90.6
BaO	—	232	202	7.24
ZnO	266	174	261	253
CdO	333	112	109	319

Table 6.
Compilation of e_{11} piezoelectric coefficients for some two-dimensional IIA/IIB-VI oxides predicted by different simulation techniques.

exploited, as opposed to where the materials only exhibit in-plane piezoelectric properties. The computational simulation with different approaches and the experimental corroboration of the piezoelectric performance of two-dimensional materials must be exhaustively studied to guarantee that the values of the piezoelectric coefficients are exploited more efficiently for all the applications that are proposed. In the previous discussion, it can be deduced that some materials do not present piezoelectric properties in bulk size; however, these are present when they are used with thicknesses of some atomic layers and thus are designated two-dimensional materials. Even the opposite of what was mentioned above is also feasible. These materials can be surface-modified appropriately under one or more crystal directions [28]. The addition of atoms and/or defects at the surface level is one of the possibilities to achieve materials with piezoelectric or ferroelectric properties. Even the free electrical charges in these materials must be controlled by either physical or virtual gates depending on the properties of the material to achieve piezoelectricity. The mechanical stability analysis of the piezoelectric behavior has allowed us to determine that the order of stability decreases in the following order: oxides, sulfides, selenides, and tellurides. Furthermore, the mechanical stability is reduced as the radius of the transition metal is decreased in the case of transition metal dichalcogenides (TMDs). To guarantee mechanical stability in a 2D piezoelectric material, it must have a low heat of formation.

A recent type of SnXY Janus monolayers (where X = Te, Se, S, O; Y = Te, Se, S, O; X ≠ Y) is being investigated because it presents static, dynamic, electronic, and thermodynamic stabilities that can be exploited to produce two-dimensional piezoelectric materials [29]. Of this family of two-dimensional materials, those based on tin behave as direct band semiconductors (SnOS and SnOSe with band gaps of 1.74 and 0.33, respectively) or indirect band semiconductors (SnSSe, with band gap of 1.69 eV). The d_{11} piezoelectric deformation coefficient of selenium tin oxide (SnOSe) reaches a value of 27.3 pm/V, which is an order higher than that reported for materials such as MoS_2 or quartz. Two-dimensional tin-based chalcogenides using Janus monolayers can be applied for piezoelectric applications such as energy harvesting and sensors.

4. Applications

The direct use of two-dimensional piezoelectric materials is in the implementation of compact sensors and actuators, flexible electronics, micro-electromechanical systems (MEMS), as well as energy harvesting that takes advantage of both the direct and inverse piezoelectric effects [11, 12]. Piezoelectric materials can be applied in the implementation of nanogenerators, information storage, and piezo-catalysis, as well as in biomedicine [3]. With the introduction of sensor networks and the Internet of Things (IoT), batches of sensors that are capable of being self-powered and operating as energy harvesters that exploit piezoelectric properties need to be researched and developed [12]. The first piezoelectric microgenerator was proposed by Glynne-Jones et al. in 2001 [25]. The nanogenerator concept was first proposed by Zhong Ling Wang et al. in 2006 [26]. After this, researchers around the world launched extensive research to develop nanomaterials and nanosystems to convert mechanical energy into electrical energy. Energy harvesters can then be considered the miniaturized replacement for battery-based power supplies for fully portable and/or wearable applications. Piezoelectric nanogenerators provide green and sustainable energy to implement self-powered nanosensors and nanosystems that operate wirelessly and in real time [23]. Nanosystems that directly benefit from self-powered systems are resonators, optoelectronic devices, and biosensors [24]. Furthermore, these nanogenerators can be used in piezo-photonics to tune the performance of photovoltaic devices and/or solar cells. Self-powered systems avoid frequent charging and replacement that are required by battery-based power systems. Battery-based systems take up too much space and are very heavy, which limits their portability and ability to incorporate them into wearable systems. Human activities such as finger typing and breathing may be capable of generating electrical powers on the order of 6.9 mW and 0.83 W, respectively. Wearable electronics require powers of the order of 200 microwatts to 1 watt, which can be achieved by the natural biomechanics of the human being, and its reduction is feasible when designing all systems using nanomaterials. Among the biomechanical movements from which energy can be harvested are the movement of elbow joints, heel strikes, leg movements, and arm swings [26].

If a deformation by tension or compression is applied to a piezoelectric material, a piezo-potential is generated in the pair of metal electrodes found at the ends of the material [3]. Electrons and holes as electric carriers are attracted to the piezo-potential with opposite polarity and an electric current is generated on a charge. If the strain is produced continuously, then a continuous current and voltage are generated. In this way, the nanogenerator converts mechanical energy into electrical energy. Nanogenerators based on piezoelectric two-dimensional materials have been implemented using molybdenum diselenide ($MoSe_2$) [12], α-phase indium selenide (α-In_2Se_3) [13], black phosphorus or phosphorene (BP) [14], molybdenum disulfide (MoS_2) [15], MoS_2/In_2Se_3 Van der Waals heterostructure [16], hexagonal boron nitride (h-BN) [10], zinc oxide (ZnO) [17], tungsten diselenide (WSe_2) [18], and other materials.

A piezoelectric nanogenerator based on a molybdenum diselenide ($MoSe_2$) nanosheet has been used to power a molybdenum disulfide (MoS_2)-based pH sensor and a photodetector based on a molybdenum disulfide-tungsten diselenide (MoS_2/WSe_2) [12]. This molybdenum diselenide ($MoSe_2$)-based nanogenerator provides an output voltage of 60 mV with a strain of 0.6%, which is approximately 50% larger than for a molybdenum disulfide (MoS_2)-based nanogenerator. Thanks to its excellent performance, this nanogenerator is capable of non-invasively monitoring vital signs to determine the respiratory rate and heart rate.

A simple boron nitride (BN) nanosheet when mechanically deformed can produce an alternating piezoelectric output of 50 mV and 30 pA [10]. For this material, a piezoelectric voltage coefficient (g_{11}) was determined experimentally with a value of

2.35×10^{-3} Vm/N. When this material was deposited as active material on a polyimide substrate, an energy harvester was produced with an output voltage of 9 V, a current of 200 nA, and an output power of 0.3 μW.

Human skin is an organ capable of perceiving external environmental stimuli or changes against variables such as temperature, humidity, and pressure [19]. Applications such as prosthetics, medical equipment, wearable devices, robots, and others have benefited from the development of electronic skins. The concept of artificial or electronic skin was proposed in the early 1980s by George Lucas as a future application concept. The first versions implemented showed limited flexibility, low resolution, and poor sensitivity. Therefore, new versions must take advantage of artificial intelligence and wearable technology for the development of health monitoring and prosthetic devices. In addition, the active materials to design these electronic skins must be sensitive, flexible, and independent of their shape and size. The application of piezoelectric nanogenerators and piezotronics allows the implementation of electronic skins that can exceed even the performance of electronic skins for the development of sensors with high spatial resolution, fast response speed, ultra-sensitivity, low power consumption, excellent durability, and ability to electrical self-supply.

IIIA-VIA compounds exhibit the coexistence of in-plane and out-of-plane piezoelectricity caused by hexagonal stacking, which makes them interesting for energy harvesting and electronic skin [13]. An α-phase indium selenide (α-In_2Se_3)-based nanogenerator was implemented with 0.76% strain producing a peak voltage and current of 35.7 mV and 47.3 pA, respectively. The d_{33} piezoelectric coefficient of α-In_2Se_3 changes from 0.34 pm/V for a monolayer to 5.6 pm/V for the bulk version.

One of the great challenges of piezoelectric materials is to produce out-of-plane polarization in active materials by exerting stress along the direction perpendicular to the nanosheet [20]. Achieving this polarization could improve the efficiency of piezoelectric transfer and make medical devices such as sphygmomanometers (for indirect blood pressure measurement) and tactical sensors such as bionic robot skins a reality. Heterostructures based on two-dimensional materials such as tin nitride (Sn_3N_4)-indium oxide (In_2O_3), germanium nitride (Ge_3N_4)-gallium oxide (Ga_2O_3), and silicon nitride (Si_3N_4)-aluminum oxide (Al_2O_3) have been studied by first-principle calculations seeking to increase the out-of-plane piezoelectric coefficients. The piezoelectric coefficients d_{33} and e_{33} were determined for the two-dimensional materials and their values have been plotted in **Figure 6**. The maximum value of d_{33} is 5290 pm/V corresponding to Sn_3N_4-In_2O_3 and the maximum value of e_{33} is 3869 pC/N corresponding to Si_3N_4-Al_2O_3. These materials are made prospects for energy harvesting applications as well as for blood pressure meters or bionic skin of robots.

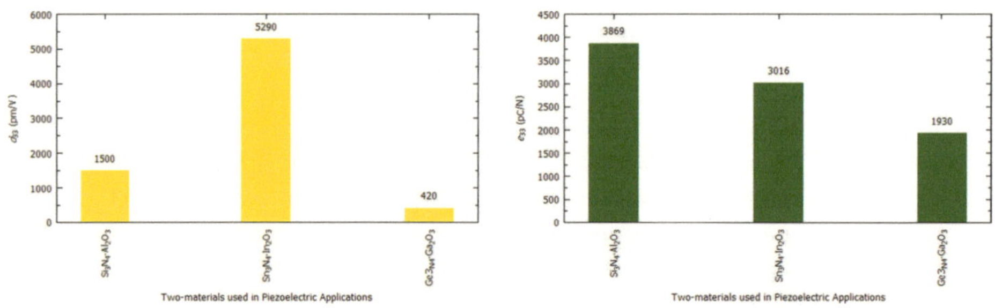

Figure 6.
d_{33} and e_{33} piezoelectric coefficients for some heterostructures based on two-dimensional materials.

Phosphorene or black phosphorus (BP) presents interesting properties such as thickness-dependent bandgap and high carrier mobility, due to its anisotropic optical, electronic, mechanical, thermal, and ionic transport characteristics [14]. Phosphorene has a d_{11} piezoelectric strain coefficient of −9.48 pm/V in the armchair direction. Phosphorene nanosheets deliver an output current of 4 pA when a compression strain of −0.72% is applied to them. With all these qualities, phosphorene can be proposed for use in strain sensors, nanogenerators, and piezoelectrically tuned transistors.

Due to a low out-of-plane piezoelectric response of α-In_2Se_3, research has been initiated on the possibility of modifying two-dimensional materials through heterostructures to increase their value. The use of the α-In_2Se_3/MoS_2 heterostructure can increase the value of the d_{33} piezoelectric coefficient [16]. The d_{33} piezoelectric coefficient for the α-In_2Se_3/MoS_2 heterostructure has a value of 17.5 pm/V. This value increases with the increase in the number of layers of the two-dimensional material but becomes saturated when it reaches 40 nm in thickness.

For a tungsten diselenide (WSe_2) bilayer nanosheet, the d_{11} piezoelectric coefficient was 3.26 ± 0.3 pm/V, whose value is higher than the d_{11} coefficient of 2.3 pm/V for volumetric α-quartz [18]. The mechanical deformation for a bilayer is 0.95%, while for a monolayer it is 0.63%. WSe_2 bilayers feature high piezoelectric coefficients and good mechanical durability for a wide strain range that could be exploited to harvest the energy required by a small liquid crystal display without applying an external power supply.

Next, a diversity of two-dimensional materials is proposed for the development of sensors and energy harvesting, and the values of the piezoelectric coefficients reached are reported, seeking their application, especially for their application outside the plane. Materials such as graphene, thanks to surface modification techniques, can produce in-plane and out-of-plane piezoelectricity reaching 37,000 pC/N for the d_{33} coefficient, 12,500,000 pm/V for the e_{33} coefficient, and 0.3 pm/V for the d_{31} coefficient [21]. For phosphorene, the values of the piezoelectric coefficients are 59 pC/N for the e_{11} coefficient and 1.06 pC/N for the e_{31} coefficient. The piezoelectric coefficient e_{11} for α-phase indium selenide (α-InSe) is 57 pC/N, and for fluorinated hexagonal boron nitride (F-h-BN) it is 84 pC/N. For phosphorene oxide (P_4O_2)-two-dimensional monolayer, the piezoelectric coefficients reach values of 54 pm/V for d_{11}, −10 pm/V for d_{31}, and 21 pm/V for d_{26}. For lithium-doped black phosphor (P_4Li_2), the piezoelectric coefficient d_{31} ranged from 2.5 to 6.28 pm/V depending on the approximation used. Two-dimensional materials such as antimonene and arsenene can reach strain limits of 58 and 24% in the armchair and zigzag directions, respectively, which are higher when compared to other two-dimensional materials such as graphene, molybdenum disulfide (MoS_2), or phosphorene. Other two-dimensional materials that offer out-of-plane piezoelectricity are oxygen-functionalized MXenes (M_2CO_2) where M can be scandium (Sc), yttrium (Y), or lanthanum (La), which have a d_{31} coefficient in the range of 0.4 to 0.78 pm/V, and a coefficient e_{31} in the range of 88 to 196 pC/N, for the case of relaxed ions [22].

The use of vertically aligned two-dimensional flexible zinc oxide nanodiscs for the design of a piezoelectric nanogenerator was reported in Ref. [23]. This nanogenerator used thermally annealed discs and generated a direct current (DC) output voltage of 17 V and a current density of 150 nA/cm^2. These values increased by 7 times the voltage and 5 times the current density if the pristine version of the same material had been used. This performance improvement was achieved thanks to superficial passivation and the reduction of oxygen vacancies in the two-dimensional material.

A recent alternative that has been reported is the possibility of developing nanogenerators with piezoelectric and triboelectric properties to develop self-powered systems [24]. Achieving the maximum performance of this nanogenerator implies taking advantage of the synergistic coupling between both types of mechanisms leading to increased electrical outputs as well as raising the energy conversion efficiency. When two materials that are electrically charged are placed in friction with each other, electrification is produced which is induced by contact, giving rise to a triboelectric effect. Like the piezoelectric effect, a triboelectric couple is produced which is directly dependent on the relative electrical polarity induced by the induced electrical charge. Two-dimensional materials such as hexagonal boron nitride (*h*-BN), metal-organic scaffolds (MOFs), and transition metal dichalcogenides (TMDs) have been proposed to induce triboelectricity. Some of these materials exhibit both triboelectric properties and piezoelectric effects. Thanks to these qualities, it is possible to decrease the internal resistance, increase the generation of electrical charge, and produce additional electrical charge trapping sites. Involving both electrical mechanisms in the same design makes it feasible to increase the total power output since both effects work efficiently under mechanical deformation, be it compression, tension, and/or vibration. The triboelectric effect operates on the surface of the material, while the piezoelectric effect operates on the volume of the material below the surface. In this way, it is possible to increase polarization by changing the structure of the material, optimizing energy harvesting, and maximizing energy conversion.

According to the previous paragraphs, the main applications derived from energy harvesting with two-dimensional piezoelectric materials are summarized in **Figure** 7. In addition to the applications summarized in **Figure** 7, there is the possibility of developing humidity, magnetic field, and mechanical force sensing. These materials possess unique piezoelectric properties relative to their nanowire-based or volumetric counterparts. Laboratory tests as well as computer simulations have shown that two-dimensional piezoelectric materials can be easily modified to achieve different piezoelectric coefficients by including a different number of layers in the design [27].

An alternative strategic implementation for harvesting mechanical energy is the development of piezoelectric nanogenerators [30]. Materials such as molybdenum disulfide (MoS_2) can take advantage of their centrosymmetric structures to produce electricity through the distortion of the crystalline lattice due to the mechanical

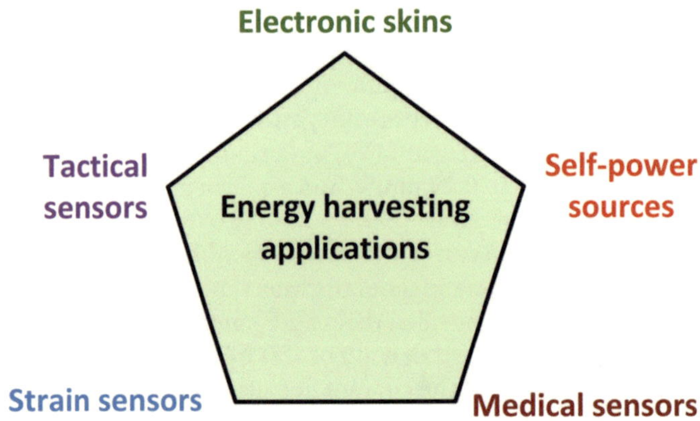

Figure 7.
Main applications from energy harvesting with two-dimensional piezoelectric materials.

deformation produced by the polarization of the charge of the constituent ions. However, an odd number of layers in the two-dimensional material structure must be used to achieve piezoelectric voltage and current outputs, which is not possible for structures with an even number of layers. Better results in piezoelectric performance are achieved when the number of layers tends to a smaller value.

Recently, a direct current generator using piezoelectric two-dimensional ZnO nanosheets has been implemented to produce an open-circuit voltage of 0.9 V, a current density for short-circuit current of 16.4 μA, and a power density of 600 nW/cm^2 for 4000 cycle operation using 4 kg of force [31]. Due to its mechanical reliability, flexibility, and high output power, this generator has the potential to be used as a power source for portable devices and as a mechanical sensor.

Due to the small thickness and light weight of two-dimensional materials, the suspended application of these materials produces significant mechanical fragility [32]. Therefore, the practical application of these materials involves the use of substrate materials to guarantee good quality of the layers as well as good stability in all the senses previously discussed. In this way, the deformation on the layers of the two-dimensional material can be homogeneous and precisely controlled throughout the substrate. Therefore, the use of polymer-based composite materials that include two-dimensional material is one of the common strategies to replace the use of a fixed and inflexible substrate, especially for wearable piezoelectric electricity generation applications.

Two-dimensional materials based on cobalt telluride ($CoTe_2$) can be used to generate electricity from waste heat using triboelectric and piezoelectric properties as energy harvesters [33]. The piezo-triboelectric nanogenerator can produce a voltage of 5 Volts when a force of 1 N is applied to it operating in a temperature range of 32 to 90 degrees Celsius.

Both mechanical flexibility and bandgap tunability are the strategic advantages of using two-dimensional materials in the implementation of data memories and electronic sensors [34]. To more appropriately exploit two-dimensional materials, it is necessary to increase the quality of their synthesis on a large scale and at a low cost, understand the relationship between the magnetic domain pattern and the applied external electric field, as well as determine the values of the piezoelectric coefficient, Curie temperature, and polarization value and in all crystalline directions of the material.

Piezoelectric materials take advantage of the mechanical deformation produced by ambient energies to develop applications such as nanogenerators as well as optical, mechanical, and magnetic sensors [35]. These make use of electrical polarization to perpendicularly deform a material by either stretching it when a positive voltage is applied or contracting it when a negative voltage is applied. Magnetic behavior is achieved when a two-dimensional material is doped or vacancies or defects are induced on the original structure.

Piezoelectric materials can replace batteries by exploiting energy from the environment for the design of self-powered devices with power consumption in the range of microwatts or milliwatts [36]. These materials offer an environmentally friendly alternative by avoiding the disposal of waste batteries that have not been manufactured and recycled with green technologies.

5. Future research directions

One of the great challenges of this century is to exploit the piezoelectric properties for the implementation of functional, sensitive, and innovative electronic devices

[21]. With the miniaturization of electronic devices, it is necessary to develop strategies for the selection of materials that can be exploited for this purpose. There are more than 7000 possible two-dimensional materials that can be modified to achieve optimization of piezoelectric properties. Among the strategies to modify the piezoelectric properties are deformation, atom or Janus substitution, functionalization, and introduction of defects in a premeditated way. Furthermore, it is possible to stack the two-dimensional nanosheets with similar or dissimilar materials to design heterostructured materials whose piezoelectric properties are completely different from those of their components. Since the piezoelectric properties thanks to these strategies can be tuned for the design of pressure sensors, piezotronics, piezo-catalysis, and energy harvesting, researchers around the world will continue to develop scientific research to take full advantage of piezoelectricity in two-dimensional materials. The suitability of the chosen material to exploit piezoelectricity comes from the following factors: difference in electronegativity between the atoms of the unit cell of the material, impact on health and the environment of the material, cost reduction and ease of the synthesis process of the material, as well as additional material properties. These factors must be considered to choose, design, and integrate the best two-dimensional materials to take the design from the laboratory phase to the practical phase for commercial production.

The study of piezoelectric two-dimensional materials is not complete [21]. Both computational modeling and experimental characterization should be further developed to predict in-plane and out-of-plane piezoelectric behaviors more accurately for various possible theoretical and technological possibilities. In addition, it is necessary to establish standards for test protocols, study the triboelectric effects involved, and complement the necessary terminology to be able to study the piezoelectric properties of two-dimensional materials. Despite the progress made in the research of two-dimensional piezoelectric materials, a lot of research must be carried out to understand their piezoelectric behavior because conventional models and theories are not able to explain the effects found in them. The use of density functional theory and molecular dynamics (MD) calculation will continue to be a vital reference source for calculating, optimizing, and predicting the piezoelectric properties of two-dimensional piezoelectric materials. A comprehensive study of the differential charge density, surface electronegativity, difference in atomic radii, anion-cation polarization ratio, effective Born charges, and elastic constants of two-dimensional piezoelectric materials must be developed to exploit the next generation of applications of high-added value.

6. Conclusions

Two-dimensional materials have undoubtedly attracted the interest of researchers around the world not only for their extraordinary properties but also for the innumerable possibilities of technological development and unprecedented scientific research. The wide range of possibilities to produce anisotropy in its piezoelectric properties extends its applications in energy harvesting, tactical sensors, medical devices, and electronic skins. Despite the advances achieved so far, computational modeling and experimental characterization of the piezoelectric properties of two-dimensional materials are still necessary to achieve a complete study of the most suitable materials to take advantage of the properties in conventional and emerging applications. The graphs of the piezoelectric coefficient values presented in this chapter illustrate, in addition to the great diversity of possible materials, a wide possibility of both in-plane and out-of-plane coefficient values. Concerning the piezoelectric properties of commonly used zinc oxide (ZnO) and gallium nitride (GaN) nanowires, the piezoelectric coefficients of the two-dimensional materials are 2 orders of magnitude larger. 2D piezoelectric materials can withstand very large deformations for their dimensions. In this century, all two-dimensional materials must be synthesized and studied to exploit the piezoelectric properties, and these can be exploited with maximum efficiency by knowing the specific conditions suitable for each material.

Acknowledgements

The author appreciates the support of the University of Guanajuato to develop this research. The author appreciates the support of the researchers who shared their publications to complement this study.

Thanks

The author wants to thank his wife and son for their support and time to edit this book. The author appreciates the support of Tea Jurcic working for IntechOpen as an author service manager.

Author details

Rafael Vargas-Bernal
University of Guanajuato, Salamanca, Guanajuato, Mexico

*Address all correspondence to: r.vargas@ugto.mx

References

[1] Li J-F. Fundamentals of piezoelectricity. In: Lead-Free Piezoelectric Materials. Weinheim, Germany: Wiley-VCH; 2021. pp. 1-18. DOI: 10.1002/9783527817047.ch1

[2] Tahir MB, Fatima U. Recent trends and emerging challenges in two-dimensional materials for energy harvesting and storage applications. Energy Storage. 2022;**4**(1):e244. DOI: 10.1002/est2.244

[3] Zhang Q, Zuo S, Chen P, Pan C. Piezotronics in two-dimensional materials. InfoMat. 2021;**3**(9):987-1007. DOI: 10.1002/inf2.12220

[4] Cheon G, Duerloo K-AN, Sendek AD, Porter C, Chen Y, Reed EJ. Data mining for new two- and one-dimensional weakly bonded solids and lattice-commensurate heterostructures. Nano Letters. 2017;**17**(3):1915-1923. DOI: 10.1021/acs.nanolett.6b05229

[5] Li R, Cheng Y, Huang W. Recent progress of Janus 2D transition metal chalcogenides: From theory to experiments. Small. 2018;**14**(45):1802091. DOI: 10.1002/smll.201802091

[6] Seixas L. Janus two-dimensional materials based on group IV monochalcogenides. Journal of Applied Physics. 2020;**128**(4):045115. DOI: 10.1063/5.0012427

[7] Blonsky MN, Zhuang HL, Singh AK, Hennig RG. Ab initio prediction of piezoelectricity in two-dimensional materials. ACS Nano. 2015;**9**(10):9885-9891. DOI: 10.1021/acsnano.5b03394

[8] Gao R, Gao Y. Piezoelectricity in two-dimensional group III-V buckled honeycomb monolayers. Physica Status Solidi: Rapid Research Letters. 2017;**11**(3):1500412. DOI: 10.1002/psrr.201600412

[9] Alyörük MM. Piezoelectric properties of monolayer II-VI group oxides by first-principles calculations. Physica Status Solidi B. 2016;**253**(12):2534-2539. DOI: 10.1002/pssb.201600387

[10] Lee G-J, Lee M-K, Park J-J, Hyeon DY, Jeong CK, Park K-I. Piezoelectric energy harvesting from two-dimensional boron nitride nanoflakes. ACS Applied Materials & Interfaces. 2019;**11**(41):37920-37926. DOI: 10.1021/acsami.9b12187

[11] Wang Y, Vu L-M, Lu T, Xu C, Liu Y, Ou JZ, et al. Piezoelectric responses of mechanically exfoliated two-dimensional SnS_2 nanosheets. ACS Applied Materials & Interfaces. 2020;**12**(46):51662-51668. DOI: 10.1021/acsami.0c16039

[12] Li P, Zhang Z. Self-powered 2D material-based pH sensor and photodetector driven by monolayer $MoSe_2$ piezoelectric nanogenerator. ACS Applied Materials & Interfaces. 2020;**12**(52):58132-58139. DOI: 10.1021/acsami.0c18028

[13] Xue F, Zhang J, Hu W, Hsu W-T, Han A, Leung S-F, et al. Multidirection piezoelectricity in mono- and multilayered hexagonal α-In_2Se_3. ACS Nano. 2018;**12**(5):4976-4983. DOI: 10.1021/acsnano.8b02152

[14] Ma W, Lu J, Wan B, Peng D, Xu Q, Hu G, et al. Piezoelectricity in multilayer black phosphorus for piezotronics and nanogenerators. Advanced Materials. 2020;**32**(7):1905795. DOI: 10.1002/adma.201905795

[15] Han JK, Kim S, Jang S, Lim YR, Kim S-W, Chang H, et al. Tunable piezoelectric nanogenerators using flexoelectricity of well-ordered hollow 2D MoS_2 shells arrays for energy harvesting. Nano Energy. 2019;**61**:471-477. DOI: 10.1016/j.nanoen.2019.05.017

[16] Yuan S, Io WF, Mao J, Chen Y, Luo X, Hao J. Enhanced piezoelectric response of layered In_2S_3/MoS_2 nanosheet-based van der Waals heterostructures. ACS Applied Nano Materials. 2020;**3**(12):11979-11986. DOI: 10.1021/acsanm.0c02513

[17] Chen J, Qiu Y, Yang D, She J, Wang Z. Improved piezoelectric performance of two-dimensional ZnO nanodisks-based flexible nanogenerators via ZnO/Spiro-MeOTAD PN junction. Journal of Materials Science: Materials in Electronics. 2020;**31**:5584-5590. DOI: 10.1007/s10854-020-03124-0

[18] Lee J-H, Park JY, Cho EB, Kim TY, Han SA, Kim T.H, Liu Y, Kim SK, Roh CJ, Yoon H-J, Ryu H, Seung W, Lee JS, Lee J, Kim S-W. Reliable piezoelectricity in bilayer WSe_2 for piezoelectric nanogenerators. Advanced Materials 2017; 29(29): 1606667. DOI: 10.1002/adma.201606667

[19] Yuan H, Lei T, Qin Y, Yang R. Flexible electronic skins based on piezoelectric nanogenerators and piezotronics. Nano Energy. 2019;**59**:84-90. DOI: 10.1016/j.nanoen.2019.01.072

[20] Zhu D-R, Wu Y, Zhang H-N, Zhu L-H, Zhao S-N. New direction's piezoelectricity and new applications of two-dimensional group V-IV-II-VI films: A theoretical study. Physica E: Low-dimensional Systems and Nanostructures. 2020;**124**:114214. DOI: 10.1016/j.physe.2020.114214

[21] Sherrell PC, Fronzi M, Shepelin NA, Corletto A, Winkler D, Ford M, et al. A bright future for engineering piezoelectric 2D crystals. Chemical Society Reviews. 2022;**51**(2):650-671. DOI: 10.1039/d1cs00844g

[22] Tan J, Wang Y, Wang Z, He J, Liu Y, Wang B, et al. Large out-of-plane piezoelectricity of oxygen functionalized MXenes for ultrathin piezoelectric cantilevers and diaphragms. Nano Energy. 2019;**65**:104058. DOI: 10.1016/j.nanoen.2019.104058

[23] Verma K, Bharti DK, Badatya S, Srivastava AK, Gupta MK. A high performance flexible two dimensional vertically aligned ZnO Nanodisc based piezoelectric nanogenerator via surface passivation. Nanoscale Advances. 2020;**2**(5):2044-2051. DOI: 10.1039/c9na00789j

[24] Zhang J, He Y, Boyer C, Kalantar-Zadeh K, Peng S, Chu D, et al. Recent developments of hybrid piezo-triboelectric nanogenerators for flexible sensors and energy harvesters. Nanoscale Advances. 2021;**3**(19):5465-5486. DOI: 10.1039/d1na00501d

[25] Glynne-Jones P, Beeby SP, White NM. Towards a piezoelectric vibration-powered microgenerator. IEE Proceedings: Science, Measurement and Technology. 2001;**148**(2):68-72. DOI: 10.1049/ip-smt:20010323

[26] Wang ZL, Song J. Piezoelectric nanogenerators based on zinc oxide nanowire arrays. Science. 2006;**312**(5771):242-246. DOI: 10.1126/science.1124005

[27] Zhang J. On the piezopotential properties of two-dimensional materials. Nano Energy. 2019;**58**:568-578. DOI: 10.1016/j.nanoen.2019.01.086

[28] Hinchet R, Khan U, Falconi C, Kim S-W. Piezoelectric properties in

two-dimensional materials: Simulations and experiments. Materials Today. 2018;**21**(6):611-630. DOI: 10.1016/j.mattod.2018.01.031

[29] Zhang X, Cui Y, Sun L, Li M, Du J, Huang Y. Stabilities, and electronic and piezoelectric properties of two-dimensional tin dichalcogenide derived Janus monolayers. Journal of Materials Chemistry C. 2019;7(42):13203-13210. DOI: 10.1039/c9tc04461b

[30] Fan FR, Wu W. Emerging devices based on two-dimensional monolayer materials for energy harvesting. Research. 2019;**2019**:7367828. DOI: 10.34133/2019/7367828

[31] Lee Y, Kim S, Kim D, Lee C, Park H, Lee J-H. Direct-current flexible piezoelectric nanogenerators based on two-dimensional ZnO nanosheet. Applied Surface Science. 2020;**509**:145328. DOI: 10.1016/j.apsusc.2020.145328

[32] Nan Y, Tan D, Shao J, Willatzen M, Wang ZL. 2D materials as effective cantilever piezoelectric nano energy harvesters. ACS Energy Letters. 2021;**6**(6):2313-2319. DOI: 10.1021/acsenergylett.1c00901

[33] Negedu SD, Tromer R, Gowda CC, Woellner CF, Olu FE, Roy AK, et al. Two-dimensional cobalt telluride as a piezo-tribogenerator. Nanoscale. 2022;**14**(21):7788-7797. DOI: 10.1039/d2nr00132b

[34] Cui C, Xue F, Hu W-J, Li L-J. Two-dimensional materials with piezoelectric and ferroelectric functionalities. NPJ 2D Materials and Applications. 2018;**2**:18. DOI: 10.1038/s41699-018-0063-5

[35] Yang S, Chen Y, Jiang C. Strain engineering of two-dimensional materials: Methods, properties, and applications. InfoMat. 2021;**3**(4):397-420. DOI: 10.1002/inf2.12177

[36] Sachdeva PK, Bera C. Theoretical design and discovery of two-dimensional materials for next-generation flexible piezotronics and energy conversion. Applied Research. 2023:e202200116. DOI: 10.1002/appl.202200116

Chapter 3

Transient Crystal Structure of Oscillating Quartz

Shinobu Aoyagi and Hiroaki Takeda

Abstract

Piezoelectric quartz oscillators are widely used to provide a stable clock signal for watches and other electric circuits. The electrically induced mechanical vibration of quartz will be caused by ionic displacements of cationic Si and anionic O sublattices against each other. However, the transient and small ionic displacements during the mechanical vibration cannot be observed by usual X-ray structure analysis. The electrically induced mechanical vibration of quartz is resonantly amplified under an alternating electric field with the resonant frequency. We have revealed the amplified lattice strain and ionic displacements in a resonantly vibrating quartz crystal under an alternating electric field by time-resolved X-ray diffraction. The details of the experiment and application of the technique to other piezoelectric oscillators are introduced in this chapter.

Keywords: quartz oscillator, piezoelectricity, time-resolved X-ray diffraction, crystal structure analysis, langasite

1. Introduction

Inorganic crystals that exhibit piezoelectricity are currently used in a wide range of industrial applications, such as oscillators, sensors, transducers, and actuators. Piezoelectricity is the property of crystals that generates an electric polarization $P_i = \Sigma_j d_{ij}\sigma_j$ (i = 1–3) proportional to an applied stress σ_j (j = 1–6). d_{ij} are the piezoelectric constants that indicate the degree of piezoelectricity. Crystals belonging to 20 of 32 crystal point groups show piezoelectricity. Piezoelectric crystals exhibit also inverse piezoelectricity that generates a strain $s_j = \Sigma_i d_{ij}E_i$ proportional to an applied electric field E_i.

The most famous and industrially important piezoelectric crystal is quartz (α-SiO_2). Quartz is a naturally abundant mineral and can be synthesized artificially by the hydrothermal method. Quartz oscillates mechanically and electrically at a stable frequency, making it widely used in oscillators to provide reference signals for various devices such as quartz watches. However, quartz has the disadvantage that it cannot be used in high-temperature environments because it undergoes a phase transition from the low-temperature α phase to the high-temperature β phase at 573°C.

The authors have been investigating the mechanism of piezoelectricity of quartz and other piezoelectric crystals and synthesizing new piezoelectric crystals such as langasite-type crystals to develop crystals with high functional piezoelectricity superior to quartz crystals. In this chapter, we introduce transient crystal structures of quartz and langasite-type crystals oscillating under an alternating electric field measured by synchrotron X-ray diffraction [1–3]. Since inverse piezoelectricity is caused by atomic displacements under electric field, measurement of atomic displacements under an electric field is essential to understand the mechanism of piezoelectricity.

Crystal structure analysis based on X-ray diffraction is a powerful tool to measure atomic displacements in crystals, but highly accurate experiments and analysis are required to measure atomic displacements under electric fields. For example, in the case of a quartz crystal with the thickness of 0.1 mm, when a 1 kV potential difference (10 MV/m in an electric field) is applied between the surfaces, the change in the Si–O bond distance (~0.16 nm = 1.6 Å) is estimated to be only about 10^{-6} nm from its piezoelectric constants. This value is two orders of magnitude less than the standard deviations of bond distances (~10^{-4} nm) determined by conventional X-ray crystal structure analysis.

2. Time-resolved X-ray diffraction under alternating electric field

In order to measure small atomic displacements in piezoelectric crystals under an electric field, the authors have developed a new method for structure analysis of piezoelectric crystals that utilizes resonance under an alternating electric field [1]. Piezoelectric crystals vibrate mechanically and electrically at a certain natural frequency when an instantaneous stress or electric field is applied. For example, the natural frequency of a common quartz oscillator used in industry is 32,768 (2^{15}) Hz. When an AC electric field with a frequency equal to this natural frequency is applied, the piezoelectric crystal resonates, producing mechanical and electrical vibrations with a particularly large amplitude. We hypothesized that the resonance under an AC electric field could amplify atomic displacements in piezoelectric crystals to a magnitude that could be measured by X-ray crystal structure analysis. The magnitude of the amplification effect in resonance depends on the quality factor (*Q*-value) of the crystal defined by 2π[(energy stored in the system)/(energy lost from the system in one period of vibration)]. The larger the *Q*-value, the smaller the energy dissipation and the greater the amplification effect. The *Q*-value of piezoelectric crystals used in resonators is particularly high, exceeding 10^6 for quartz crystals.

Even if the atomic displacements involved in piezoelectricity can be greatly amplified by resonance under an alternating electric field, it is actually impossible to measure them using conventional X-ray diffraction. Conventional X-ray diffraction measures X-ray diffraction images during X-ray irradiation for several seconds to several minutes while the crystal is rotating, so it is impossible to measure instantaneous X-ray diffraction images of piezoelectric crystals vibrating at frequencies from kHz to MHz. Synchrotron radiation (SR) X-rays with high brilliance and short pulse duration are useful for the measurement of instantaneous X-ray diffraction images. SR is an electromagnetic wave emitted tangentially to the trajectory of an electron bunch accelerated to nearly the speed of light when the trajectory is bent by a magnetic field. We have been conducting SR time-resolved X-ray diffraction experiments under an AC electric field at SPring-8, the large SR facility (Hyogo, Japan), to measure transient atomic displacements in piezoelectric crystals resonating under an AC electric field. The shortest pulse duration is about 50 ps. By repeatedly irradiating a piezoelectric crystal resonating under an AC electric field with highly brilliant and

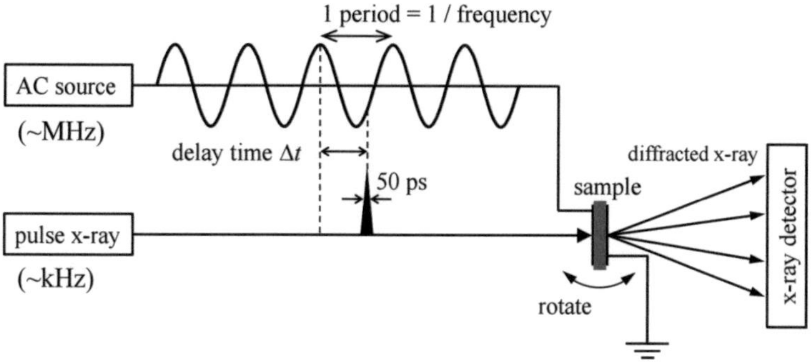

Figure 1.
Schematic of time-resolved X-ray diffraction under AC electric field.

short pulse X-rays synchronized with the AC electric field, instantaneous x-ray diffraction images when the atomic displacements reach the maximum can be measured with high accuracy (**Figure 1**).

3. Transient atomic displacements in quartz oscillator

Transient atomic displacements in a quartz oscillator were successfully measured by the SR time-resolved X-ray diffraction under an AC electric field [1]. Quartz crystal belongs to the trigonal crystal system with the point group 32. There are two types of crystal polymorphs in quartz: right and left quartz, which are enantiomorphs of each other. The industrially used quartz crystal is a right crystal with the space group $P3_221$. The crystal structure of the right quartz crystal is shown in **Figure 2a**. The crystal structure of quartz consists of corner shared SiO_4 tetrahedra with the Si–O bond distance of 1.61 Å, the O–Si–O bond angle of 109 degrees, and the Si–O–Si bond angle of 143 degrees. The direction along the twofold axis is called the *X*-axis, the direction perpendicular to the twofold and threefold axes is called the *Y*-axis, and the direction along the threefold axis is called the *Z*-axis. The *X*, *Y*, and *Z*-axes correspond to the [100], [120], and [001] crystal band axes of the primitive trigonal lattice, respectively. The nonzero piezoelectric constants of the right quartz are $d_{11} = -d_{12} = -d_{26}/2 = -2.31$ pC•N^{-1}, and $d_{14} = -d_{25} = -0.727$ pC•N^{-1} [4]. The right

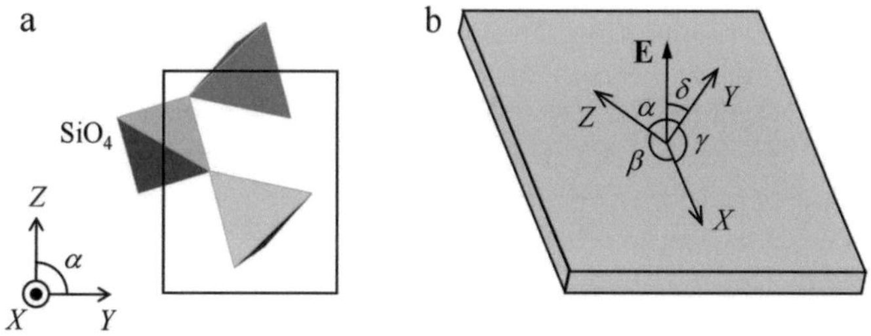

Figure 2.
(a) Crystal structure of quartz viewed along the twofold axis (square box is the trigonal unit cell), and (b) schematic of AT-cut quartz oscillator.

quartz and left are piezoelectrically distorted under the electric field E_1 along the X-axis and E_2 along the Y-axis, but not under the electric field E_3 along the Z-axis.

A commercially available AT-cut quartz oscillator with an oscillation frequency of 30 MHz was used as the sample for the measurement. AT-cut oscillators are plate-shaped crystal (**Figure 2b**) cut along the plane including the X-axis, which is perpendicular to the direction tilted from the Y-axis to the Z-axis by $\delta = 35$ degrees. It is widely used in industry because of its small temperature variation of oscillation frequency near room temperature. When no electric field is applied, the X, Y, and Z-axes are orthogonal to each other, but when an electric field E is applied along the direction perpendicular to the AT-cut plane, the shear strain given by $s_5 = d_{25}E_2$ and $s_6 = d_{26}E_2$ causes a distortion of the angle between the X and Z-axes (β angle) and angle between the X and Y-axes (γ angle) from 90 degrees (**Figure 2b**). Here, $E_2 = E\cos\delta$. The angle between the Y and Z-axes (α angle) is not distorted under the electric field. The oscillation frequency of an AT-cut oscillator f_0 is given by $f_0 = 1664/h$ (MHz) where h (μm) is the thickness of the crystal. To reduce X-ray absorption and extinction effects, a thin and high frequency AT-cut quartz oscillator with $h = 55$ μm ($f_0 = 30$ MHz) was used as a sample.

X-ray diffraction experiments were performed at the SPring-8 beamline BL02B1 [5] using X-rays with a wavelength of 0.4 Å. A large cylindrical curved imaging plate was used as the X-ray detector. First, measurements under a DC electric field were performed. Piezoelectric distortions of β and γ angles from 90 degrees under a DC electric field of $E = 36$ MV•m^{-1} (2.0 kV in potential difference) were −0.002(2) and −0.007(3) degrees, respectively. These values are consistent with the values of −0.001 and −0.008 degrees calculated from the piezoelectric constants. Crystal structures determined from more than 4000 Bragg reflection intensities measured under DC electric fields of $E = +36$ and −36 MV•m^{-1} were compared. However, no differences in the bond distances and angles exceeding their standard deviations were observed.

Time-resolved X-ray diffraction of a resonantly vibrating AT-cut quartz oscillator (**Figure 1**) was performed by applying a sinusoidal AC electric field with a frequency of 30 MHz and an electric field amplitude of 0.18 MV•m^{-1} to the sample. Resonance of the sample was confirmed by detecting the current flowing in the circuit with a current probe and displaying it on an oscilloscope. The resonant sample was irradiated with pulsed X-rays with a pulse duration of ~50 ps at a repetition rate of 26 kHz using an X-ray chopper [6]. In order to synchronize the oscillation of the sample and the pulsed X-rays, the ratio of the resonance frequency of the sample to the repetition frequency of the X-rays must be an exact integer ratio. However, both of the resonance frequency of the sample and the repetition frequency of the X-rays cannot be tuned freely. The authors have synchronized the resonance of the sample with pulsed X-rays by modulating the 30 MHz AC electric field at 26 kHz. This method enables instantaneous measurement of X-ray diffraction images of a resonant sample with a time resolution of less than 1 ns. The period of the resonant sample is 33 ns = 1/30 MHz. By varying the delay time Δt of the alternating electric field to the pulsed X-rays from 0 to 33 ns, the time variation in one period of the X-ray diffraction images of the resonant sample was measured.

Figure 3a shows time dependences of deviations of α, β, and γ lattice angles from 90 degrees ($\Delta\alpha$, $\Delta\beta$, and $\Delta\gamma$) obtained by the least-squares method from the positions of several hundred Bragg reflections measured on the X-ray diffraction images. It can be seen that, during the resonance, α angle remains almost unchanged with time, while β and γ angles oscillate sinusoidally and significantly. The amplitudes of the $\Delta\beta$ and $\Delta\gamma$

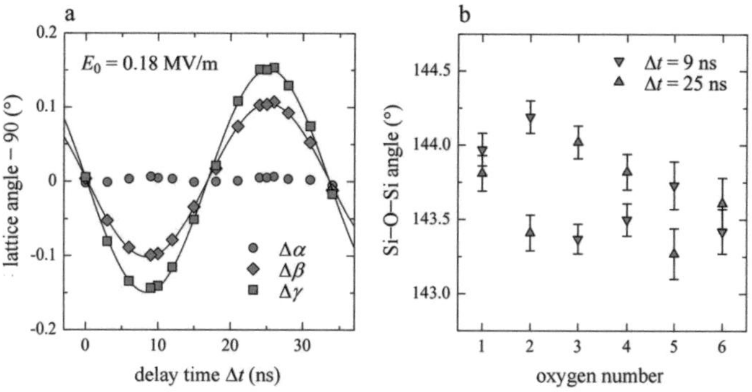

Figure 3.
(a) Time variation of Δα, Δβ, and Δγ, and (b) Si–O–Si bond angles of AT-cut quartz crystal under AC electric field.

oscillations are 0.10 and 0.15 degrees, respectively. These values are the several tens of times larger than the aforementioned values of −0.002(2) and −0.007(3) degrees under the DC electric field of 36 MV•m^{-1}. $\Delta\beta$ and $\Delta\gamma$ under a DC electric field of 0.18 MV•m^{-1} calculated from the piezoelectric constants are -1×10^{-5} and -4×10^{-5} degrees, respectively. Thus, the resonance under an AC electric field amplifies $\Delta\beta$ and $\Delta\gamma$ by a factor of 1×10^4 and 4×10^3, respectively.

The crystal structures at the delay times Δt = 9 and 25 ns when $\Delta\beta$ and $\Delta\gamma$ reach the negative and positive maxima, respectively, were obtained by the least-squares method from more than 3000 Bragg reflection intensities. Comparison of the two crystal structures shows that the Si–O bond distances and O–Si–O bond angles in the SiO_4 tetrahedra do not differ by more than their standard deviations (0.002 Å for the bond distances and 0.1 degrees for the bond angles). However, a clear difference was observed in the Si–O–Si bond angles between the SiO_4 tetrahedra. The triclinic crystal structure of the quartz crystal distorted under an electric field consists of three independent SiO_4 tetrahedra with six independent oxygen atoms (O(1)~O(6)). The six independent Si–O–Si bond angles at Δt = 9 and 25 ns are shown in **Figure 3b**. The abscissa is the number of independent oxygen atoms 1~6. Among O(1)~O(6), large deformations were observed in the Si–O–Si bond angles, especially around O(2) and O(3).

The large deformation of the Si–O–Si bond angles around O(2) and O(3) can be understood from the displacements of the anionic O atoms under the electric field and the arrangements of the Si–O–Si bonds relative to the electric field. The Si–O bonds have both covalent and ionic characters. The electron density distribution was calculated from the measured X-ray diffraction intensities to estimate the numbers of electrons of each atom. The charge deviations from neutral were +2.8*e* for Si and −1.4*e* for O (*e* is the elementary charge). The Si and O atoms, which are positively and negatively charged, respectively, are expected to be displaced in opposite directions upon application of an electric field. However, each Si atom is covalently bonded to four O atoms in tetrahedral coordination, which makes it extremely difficult to displace the Si atoms by an electric field. O atoms are subject to a large repulsive force due to the Si–O covalent bonds when the displacement direction is parallel to the Si–O–Si plane, but can be displaced relatively easily when the displacement direction is perpendicular to the Si–O–Si plane. Comparing the angles between the electric field and the Si–O–Si planes of O(1)~O(6), O(2), and O(3) with the angles about

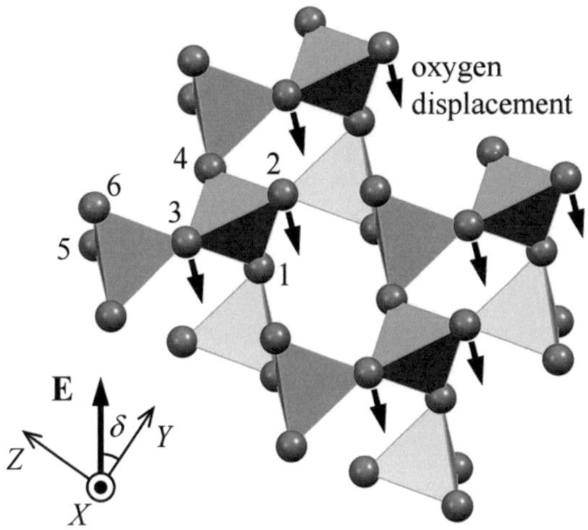

Figure 4.
Oxygen atomic displacements in AT-cut quartz crystal viewed along the twofold axis under electric field **E** *perpendicular to the crystal plane.*

66 degrees are the closest to perpendicular. The Si–O–Si bond angles centered on O(2) and O(3) deform with the oxygen displacements in the direction perpendicular to the Si–O–Si planes. The relationship between the crystal structure, electric field, and atomic displacements of O(2) and O(3) is shown in **Figure 4**.

As shown above, the piezoelectric distortion of a quartz crystal under an electric field is caused by the displacement of oxygen ions in the direction perpendicular to the Si–O–Si plane due to the electric field and the accompanying deformation of the Si–O–Si bond angles. Mechanical and electrical vibrations with high *Q*-values of quartz crystals are caused by the restoring force acting on the Si–O–Si bond angles, which causes a harmonic vibration of the oxygen ions with little damping. The combination of resonant vibration under an alternating electric field and SR time-resolved X-ray diffraction has revealed the transient atomic displacements and mechanism of piezoelectricity in quartz. Using this experimental technique, the authors next performed similar transient atomic displacement measurements under an AC electric field on langasite-type crystals, which are useful as piezoelectric materials for high-temperature applications [2, 3].

4. Applications to langasite-type crystals

Langasite ($La_3Ga_5SiO_{14}$, LGS) is a piezoelectric crystal belonging to the same crystal point group 32 as quartz and does not show a phase transition until its melting point around 1470°C. Its piezoelectric constants are several times larger than those of quartz, $d_{11} = -5.95$, $d_{14} = 5.38$ pC•N^{-1} [7]. **Figure 5a** shows the crystal structure of langasite. The space group is *P*321, and the crystal structure consists of GaO_6 octahedra, GaO_4, and $Ga_{1/2}Si_{1/2}O_4$ tetrahedra, and La atoms coordinated by eight oxygen atoms. La atoms can be substituted with other rare earth (RE) elements such as Pr and Nd, and it is known that the piezoelectric constants decrease as the ionic radius of the RE elements decreases [7–10]. In $Nd_3Ga_5SiO_{14}$ (NGS), where La is replaced by Nd, $d_{11} = -4.05$, $d_{14} = 2.07$ pC•N^{-1} [7]. Transient atomic displacement measurements under an AC electric field were performed on LGS and NGS to understand the origin

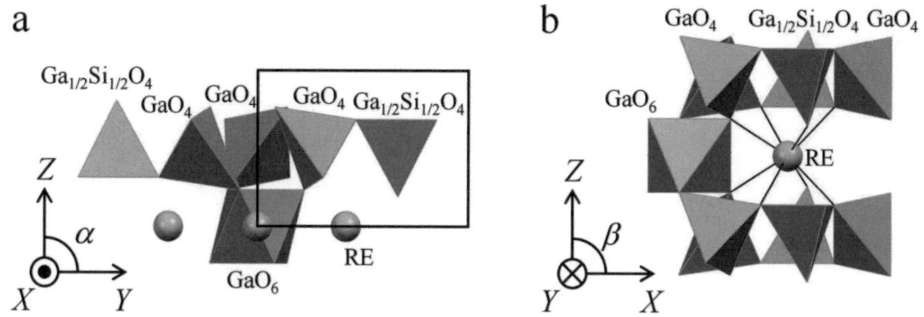

Figure 5.
(a) Crystal structure of langasite-type crystal viewed along the twofold axis (square box is the trigonal unit cell), and (b) coordination structure around a RE atom.

of the change in piezoelectric constants due to RE element substitution, in addition to the mechanism of piezoelectricity in langasite-type crystals. The samples used were a commercially available *Y*-cut LGS crystal with an oscillation frequency of 28 MHz (thickness: 0.05 mm) and a homemade *Y*-cut NGS crystal with an oscillation frequency of 13 MHz (thickness: 0.09 mm). X-ray diffraction experiments were performed at SPring-8 BL02B1 using X-rays with a wavelength of 0.3 Å.

The crystal structures of LGS and NGS in the absence of an electric field were analyzed and compared first. There was little difference in the Ga(Si)–O bond distances between them. The RE (La, Nd) atoms are coordinated with eight O atoms, but because the RE atoms are on the twofold *X*-axis, there are only four independent RE–O bonds (**Figure 5b**). Among the four types of RE–O bonds, large changes in bond lengths were observed in the two short RE–O bonds due to elemental substitutions. The bond lengths of these two RE–O bonds are 2.355 and 2.508 Å for LGS and 2.301 and 2.431 Å for NGS, respectively, indicating that the substitution of La with Nd shortened the bond distance by more than 0.05 Å. In accordance with the shortening of the bond distances, the Ga–O–Ga bond angles centered on the oxygen atoms of these two types of RE–O bonds were also significantly deformed. The two Ga–O–Ga bonds are the only Ga–O–Ga bonds existing in the crystal structure, which bridging a GaO_6 octahedron and a GaO_4 tetrahedron, and a GaO_4 tetrahedron and a $Ga_{1/2}Si_{1/2}O_4$ tetrahedron, respectively. The bond angles of these two types of Ga–O–Ga(Si) bonds are 114.2 and 122.3 degrees for LGS and 112.6 and 120.3 degrees for NGS, indicating a decrease in the bond angles of about 2 degrees with the substitution of La with Nd. In addition, NGS has a larger deformation of the GaO_6 octahedron from the regular octahedron than LGS. The O–Ga–O bond angles in the GaO_6 octahedron are 90 and 180 degrees when the GaO_6 octahedron is a regular octahedron. The maximum deviations of the O–Ga–O bond angles from 90 and 180 degrees in the GaO_6 octahedron are 12.9 and 13.8 degrees for LGS and 15.2 and 16.2 degrees for NGS.

X-ray diffraction experiments under a DC electric field showed that the lattice deformations of $\Delta\beta$ and $\Delta\gamma$ were consistent with the piezoelectric constants, as in the case of quartz. Piezoelectric constants estimated from $\Delta\beta$ and $\Delta\gamma$ were $d_{11} = -5.7$ and $d_{14} = 5.3$ pC•N^{-1} for LGS and $d_{11} = -3.8$ and $d_{14} = 2.3$ pC•N^{-1} for NGS. Time-resolved X-ray diffraction experiments under an AC electric field were performed with a sinusoidal AC electric field applied to the sample at the resonant frequency, as in the case of the quartz crystal. The electric field amplitudes were 0.20 MV•m^{-1} for LGS and 0.16 MV•m^{-1} for NGS. The repetition frequency of the pulsed X-rays was 52 kHz for LGS and 69 kHz for NGS. As in the case of the quartz crystal, the time variation of the

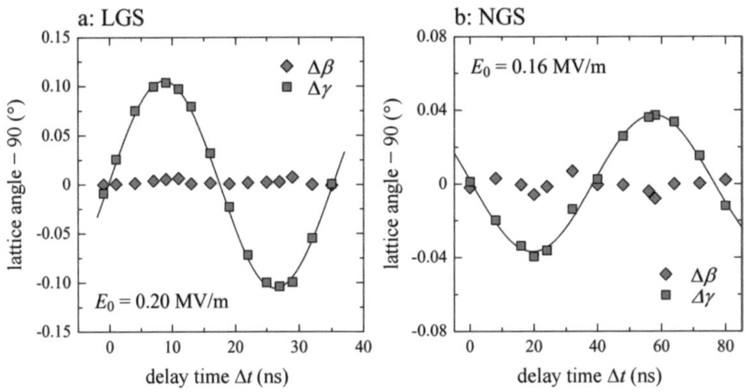

Figure 6.
Time variations of Δβ and Δγ for the (a) LGS and (b) NGS oscillators resonating under AC electric field.

X-ray diffraction image of the resonant sample in one period was measured by changing the delay time of the AC electric field to the pulsed X-rays, and the time variation of the lattice constant and crystal structure were investigated from these images.

Figure 6a and **b** show the time variation of $\Delta\beta$ and $\Delta\gamma$ for the LGS and NGS oscillators resonating under an AC electric field. As in the case of the quartz crystal (**Figure 3a**), a large sinusoidal oscillation was observed in $\Delta\gamma$. The amplitudes of the oscillation in $\Delta\gamma$ were 0.10 degrees for the LGS and 0.04 degrees for the NGS, respectively. The difference between the two values can be attributed mainly to the difference in piezoelectric constants as described above. The amplification factor of $\Delta\gamma$ under an AC electric field due to resonance is more than 5×10^2 from their piezoelectric constants, but smaller than that of a quartz crystal around 1×10^4. The larger amplification effect in quartz is attributed to its *Q*-value larger than those of LGS and NGS. The amplification effect of $\Delta\beta$ due to resonance was observed in the AT-cut quartz crystal (**Figure 3a**), but not in the *Y*-cut LGS and NGS crystals (**Figure 6**). An AC electric field was applied along the direction 35 degrees inclined from the *Y*-axis to the *Z*-axis in the AT-cut quartz crystal (**Figure 2b**), while it was applied along the *Y*-axis in the *Y*-cut LGS and NGS crystals. $\Delta\beta$ is generated by the atomic displacements perpendicular to the *Y*-axis, so an electric field component perpendicular to the *Y*-axis will be necessary to amplify $\Delta\beta$.

The crystal structures of LGS and NGS resonating under the AC electric field at the delay times when $\Delta\gamma$ reached their negative and positive maxima were analyzed and compared. Changes in RE–O bond distances, Ga–O–Ga, and O–Ga–O bond angles from the negative to positive maxima of $\Delta\gamma$ are shown in **Figures 7** and **8**. No changes in the Ga(Si)–O bond distances were observed during resonance for both LGS and NGS. In the triclinic distorted structure with 24 independent RE–O bonds under an electric field, three RE–O bonds in LGS and one RE–O bond in NGS showed a change in their bond distances of more than ±0.01 Å with time (**Figure 7a**). Among the four types of independent RE–O bonds in the trigonal non-distorted structure, only the longest RE–O bonds with the bond distances longer than 2.85 Å showed a clear time variation in the resonance. The two short RE–O bonds, which vary greatly with RE atomic substitution, show no time variation in the resonance. Focusing on the Ga–O–Ga bond angles, we observed that among the 12 types of Ga–O–Ga(Si) bonds in the distorted state under an electric field, four types of bonds

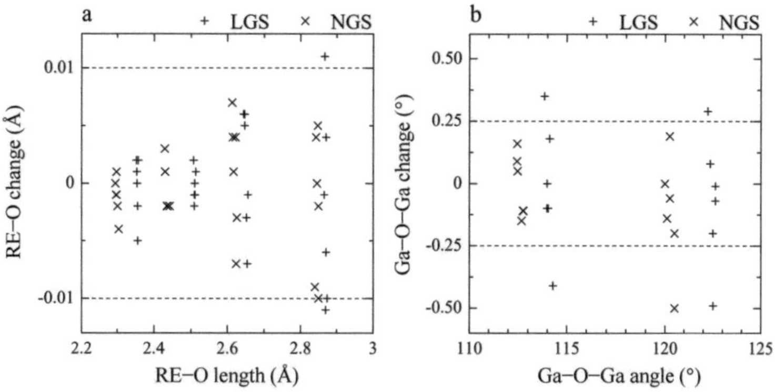

Figure 7.
Changes in (a) RE–O lengths and (b) Ga–O–Ga angles under AC electric field for Y-cut LGS and NGS resonators.

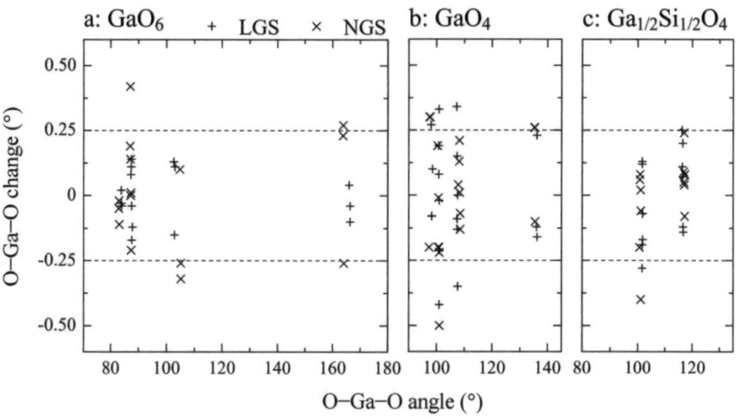

Figure 8.
Changes in O–Ga–O angles in (a) GaO_6 octahedra, (b) GaO_4, and (c) $Ga_{1/2}Si_{1/2}O_4$ tetrahedra under AC electric field for Y-cut LGS and NGS resonators.

in the LGS and one type of bond in the NGS showed a change in their bond angles of more than ±0.25 degrees with time (**Figure 7b**). These results indicate that replacing La with Nd significantly reduces the deformation of the RE–O bond distances and the Ga–O–Ga bond angles during resonance as RE–O bond distances are shortened. The decrease in the piezoelectric constants due to the replacement of La with Nd is mainly attributed to the fact that the deformation of the Ga–O–Ga bond angles is prevented due to the shortening of the RE–O bond distances.

Next, the deformation of O–Ga–O bond angles in GaO_6 octahedra, GaO_4, and $Ga_{1/2}Si_{1/2}O_4$ tetrahedra in the resonance state was investigated. Of the 15 independent O–Ga–O bonds in the GaO_6 octahedra of the distorted structure under an electric field, no and 5 O–Ga–O bonds showed a time variation of their bond angles greater than ±0.25 degrees in LGS and NGS, respectively (**Figure 8a**). In the GaO_4 tetrahedra, of the 18 independent O–Ga–O bonds, 5 and 5 O–Ga–O bonds showed a time variation greater than ±0.25 degrees in LGS and NGS, respectively (**Figure 8b**). In the $Ga_{1/2}Si_{1/2}O_4$ tetrahedra, of the 12 independent O–Ga–O bonds, 2 and 1 O–Ga–O bonds showed a time variation greater than ±0.25 degrees in LGS and NGS, respectively (**Figure 8c**). The substitution of La with Nd increases the deformation of the GaO_6 octahedra from the regular octahedron and facilitates the deformation of the O–Ga–O bond angles of the GaO_6 octahedra during resonance. The Ga–O–Ga bond angles in NGS are not easily deformed due to the shortening of RE–O bonds, but the GaO_6 octahedra are deformed instead. The deformation of GaO_4 and $Ga_{1/2}Si_{1/2}O_4$ tetrahedra during resonance was observed in both LGS and NGS, but no deformation of SiO_4 tetrahedra during resonance was observed in quartz. Therefore, the O–Ga–O bond is more flexible than the O–Si–O bond, and as a result, the piezoelectric constants of LGS and NGS are larger than those of quartz. However, the piezoelectric deformation of LGS and NGS is caused by the deformation of multiple bonds with different force constants, resulting in greater energy dissipation and lower amplification effect (Q-value) in resonance than those of quartz.

As described above, the mechanism of piezoelectricity and the effect of RE substitution in langasite-type crystals were understood by the combination of resonance phenomena under an AC electric field and SR time-resolved X-ray diffraction. LGS shows larger piezoelectric deformation than quartz due to the deformation of GaO_4 and $Ga_{1/2}Si_{1/2}O_4$ tetrahedra. In NGS, the piezoelectric deformation is smaller because the shortening of the RE–O bond distances prevents the deformation of the Ga–O–Ga bond angles, but instead the GaO_6 octahedral deformation is observed. These findings will be useful for the design and development of piezoelectric crystals with higher functionality than quartz.

5. Summary

In this chapter, we introduce transient crystal structures of quartz and langasite-type crystals resonating under an AC electric field observed by SR time-resolved X-ray diffraction. The method can detect small transient atomic displacements in piezoelectric crystals by resonantly amplifying them under an alternating electric field by a factor of 10^3–10^4. We will use this method for structural analyses of other piezoelectric crystals and expand its range of application.

Acknowledgements

This work was supported by a Grant-in-Aid for Scientific Research from the Japan Society for the Promotion of Science (JSPS) (Grants Nos. JP22H02162, JP19H02797, JP16K05017, and JP26870491), Tatematsu Foundation, Toyoaki Scholarship Foundation, Daiko Foundation, and the Research Equipment Sharing Center at the Nagoya City University. The synchrotron radiation experiments were performed at SPring-8 with the approval of the Japan Synchrotron Radiation Research Institute (JASRI).

Conflict of interest

The authors declare no conflict of interest.

Author details

Shinobu Aoyagi[1*] and Hiroaki Takeda[2]

1 Department of Information and Basic Science, Graduate School of Science, Nagoya City University, Nagoya, Japan

2 Faculty of Engineering, Department of Applied Chemistry, Saitama University, Saitama, Japan

*Address all correspondence to: aoyagi@nsc.nagoya-cu.ac.jp

References

[1] Aoyagi S, Osawa H, Sugimoto K, Fujiwara A, Takeda S, Moriyoshi C, et al. Atomic motion of resonantly vibrating quartz crystal visualized by time-resolved X-ray diffraction. Applied Physics Letters. 2015;**107**:201905. DOI: 10.1063/1.4935591

[2] Aoyagi S, Osawa H, Sugimoto K, Takeda S, Moriyoshi C, Kuroiwa Y. Time-resolved crystal structure analysis of resonantly vibrating langasite oscillator. Japanese Journal of Applied Physics. 2016;**55**:10TC05. DOI: 10.7567/JJAP.55.10TC05

[3] Aoyagi S, Osawa H, Sugimoto K, Nakahira Y, Moriyoshi C, Kuroiwa Y, et al. Time-resolved structure analysis of piezoelectric crystals by X-ray diffraction under alternating electric field. Japanese Journal of Applied Physics. 2018;**57**:11UB06. DOI: 10.7567/JJAP.57.11UB06

[4] Bechmann R. Elastic and piezoelectric constants of alpha-quartz. Physical Review. 1958;**110**:1060-1061. DOI: 10.1103/PhysRev.110.1060

[5] Sugimoto K, Ohsumi H, Aoyagi S, Nishibori E, Moriyoshi C, Kuroiwa Y, et al. Extremely high resolution single crystal diffractometery for orbital resolution using high energy synchrotron radiation at SPring-8. AIP Conference Proceedings. 2010;**1234**:887-890. DOI: 10.1063/1.3463359

[6] Osawa H, Kudo T, Kimura S. Development of high-repetition-rate X-ray chopper system for time-resolved measurements with synchrotron radiation. Japanese Journal of Applied Physics. 2017;**56**:048001. DOI: 10.7567/JJAP.56.048001

[7] Takeda H, Izukawa S, Shimizu H, Nishida T, Okamura S, Shiosaki T. Growth, structure and piezoelectric properties of $Ln_3Ga_5SiO_{14}$ (Ln = La, Nd) single crystals. Transactions of the Materials Research Society of Japan. 2005;**30**:63-66

[8] Sato J, Takeda H, Morikoshi H, Shimamura K, Rudolph P, Fukuda T. Czochralski growth of $RE_3Ga_5SiO_{14}$ (RE = La, Pr, Nd) single crystals for the analysis of the influence of rare earth substitution on piezoelectricity. Journal of Crystal Growth. 1998;**191**:746-753. DOI: 10.1016/S0022-0248(98)00362-5

[9] Iwataki T, Ohsato H, Tanaka K, Morikoshi H, Sato J, Kawasaki K. Mechanism of the piezoelectricity of langasite based on the crystal structures. Journal of the European Ceramic Society. 2001;**21**:1409-1412. DOI: 10.1016/S0955-2219(01)00029-2

[10] Araki N, Ohsato H, Kakimoto K, Kuribayashi T, Kudoh Y, Morikoshi H. Origin of piezoelectricity for langasite $A_3Ga_5SiO_{14}$ (A = La and Nd) under high pressure. Journal of the European Ceramic Society. 2007;**27**:4099-4102. DOI: 10.1016/j.jeurceramsoc.2007.02.177

Chapter 4

Magnetoelectric Composites-Based Energy Harvesters

Tarun Garg and Lickmichand M. Goyal

Abstract

Electrical energy generation from renewable resources has been a quest in the last few decades to meet the energy demand of electrical appliances and gadgets. More importantly, portable gadgets and devices, wireless sensors, etc., which rely on batteries require intermittent charging, and it is difficult to find an omnipresent continuous electrical energy source connected to a power station for these batteries. Alternate to these power stations connected to electrical energy sources is harvesting the energy from omnipresent mechanical and acoustic vibrations and AC magnetic field. Energy harvesting from these waste energy resources is possible using piezoelectric and magnetoelectric materials. This chapter would discuss in detail various mechanisms and stimuli, which may be synergistically used to harvest energy from piezoelectric materials-based energy harvesters.

Keywords: magnetoelectric, multiferroic, energy harvesting, piezoelectric, ferromagnetic

1. Introduction

Harvesting electrical energy from various wasted forms of energy in the environment could be a way to develop sustainable energy sources. However, to do so, we need to develop smart materials and structures, which could convert various wasted forms of energy into electrical energy. These smart materials and structures may enable us to develop sustainable energy sources required for powering up low-power electronic devices [1, 2]. In the recent past, a lot of research attention has been devoted to converting omnipresent mechanical and acoustic vibration energy to electrical energy using electromagnetic [3–5], electrostatic [6–8], and piezoelectric [9–11] transductions. The most popular technique among the methods is energy harvesting using the piezoelectric effect, which is proved to be advantageous over other methods due to its easier execution and higher power density. These piezoelectric energy harvesters are capable of harvesting energy from various kinds of vibration sources involving translational and rotary motions [12–14], wind [15, 16], or some fluid flow induced vibrations [17, 18]. Another ubiquitous form of wasted energy is magnetic energy found around the electrical current-carrying wires and appliances. This energy can be harvested by using magnetoelectric (ME) composites. An ME composite comprises a ferromagnetic (FM) material and a piezoelectric material combined with various phase connectivity schemes. On application of a magnetic field across such a composite, FM material undergoes magnetostriction, which leads to an elastic strain in the material. This elastic strain is transferred to the connected piezoelectric phase ***via*** a mechanical coupling. Due to this strain in the piezoelectric material, an electric field is generated in a transverse direction across the piezoelectric phase. This electric field is harvested in the form of electric energy using these ME composites. In other words, the application of a magnetic field across these composites produces an electric field. This effect is termed as the direct magnetoelectric effect. The converse effect is also true in these ME composites, where the application of an electric field leads to the generation of a magnetic field. Both effects have been explored in numerous applications such as energy harvesting, magnetoelectric transformers, AC and DC magnetic field sensors, phase shifters, and resonators. [19]. However, in this chapter, we will be discussing the direct magnetoelectric effect in ME composites, which is useful for energy harvesting. This chapter is organized in the following way: We will begin with a discussion of the magnetoelectric effect in magnetoelectric composites, the materials requirements, and the properties of the ME effect. A list of candidate materials for ME composites has been provided to give the idea of specific materials to the readers. Then, applications of these ME composites, specifically energy harvesting, have been discussed.

2. Magnetoelectric effect

ME effect is the induction of electric polarization in a material under an applied external magnetic field or conversely, the generation of magnetization due to an applied external electric field [20]. This induced polarization P depends on the applied external magnetic field H, according to the following expression:

$$\boldsymbol{P} = \alpha \boldsymbol{H} \tag{1}$$

where α is called second rank ME-susceptibility tensor.

ME effect is quantified by calculating the ME voltage coefficient α_E using the following expression:

$$\alpha_E = \frac{\alpha}{\varepsilon_o \varepsilon_r} = \left(\frac{\partial E}{\partial H}\right)_S \tag{2}$$

Here, ε_o and ε_r are the electric permittivity of free space and the relative permittivity of the dielectric material, respectively. The ME effect was first realized in single-phase materials, the so-called multiferroics, possessing both ferromagnetic and ferroelectric orderings. ME effect is a consequence of the coupling between these ordered parameters. Landau's theory describes this coupling by the free energy F of the system expressed in terms of applied field, that is, magnetic field H or electric field E [21]. The ME effect was theoretically predicted by Curie in 1894 [22]. However, it was first observed by the Russian scientist Astrov in 1960 in a single-phase material Cr_2O_3, which is antiferromagnetic [23]. Afterward, many more single-phase materials were found to show this effect. ME effect in these single-phase materials was found to be weak and rare due to the nonsimultaneous presence of two ferroic orders at the same temperature. Therefore, artificially engineered multiferroic magnetoelectric composites came into realization. The first bulk ME composite was reported by Van den Boomgaard in the 1970s. Ferroelectric $BaTiO_3$ combined with cobalt ferrite/nickel ferrite to form bulk ME composite produced a ME voltage coefficient, α_E of

130 mV/cm Oe. The constituent materials were synthesized using a solid-state reaction technique [24]. In recent years, different forms of ME composites, such as layered stacks of piezoelectric/magnetostrictive materials, polymer-ceramic matrix composites, and rare earth elements-based composites, which have shown relatively stronger intrinsic ME effect have found great research interest for applications in futuristic electronic devices.

ME effect in the composites of ferromagnetic and piezoelectric materials appears as a product tensor property, which was first proposed by van Suchtelen in 1972. None of the constituent materials used to form a composite shows the ME effect individually. However, on forming a composite, a relatively stronger ME effect results. In this composite, the mechanical deformation in ferromagnetic material due to magnetostriction results in an electrical polarization due to the piezoelectric effect in the piezoelectric material. The product of the magnetostrictive effect (magnetic/mechanical effect) in ferromagnetic material and the piezoelectric effect (mechanical/electrical effect) in piezoelectric material gives a direct ME effect [25]:

$$\text{Direct ME effect} = \text{magnetic/mechanical}_{\times}\ \text{mechanical/electric} \tag{3}$$

$$\text{Converse ME effect} = \text{electric/mechanical}_{\times}\ \text{mechanical/magnetic} \tag{4}$$

A schematic for the ME effect in composites is shown in **Figure 1**. Direct ME effect is observed when the ME composite is subjected to a magnetic field, which causes a change in the shape of the ferromagnetic phase due to magnetostriction. In other words, a strain is developed in ferromagnetic material due to the application of a magnetic field. This strain is then transferred ***via*** mechanical coupling to the adjacent piezoelectric phase, causing an electric polarization in it due to the piezoelectric effect.

For the ferromagnetic phase, due to the application of a magnetic field to the ME composite,

$$\left(\frac{\partial \boldsymbol{S}}{\partial \boldsymbol{H}}\right)_{\boldsymbol{T}} = \boldsymbol{q} \tag{5}$$

and for the piezoelectric phase

$$\left(\frac{\partial \boldsymbol{P}}{\partial \boldsymbol{S}}\right)_{\boldsymbol{E}} = \boldsymbol{d} \tag{6}$$

where S is the strain, T is stress, and q (where $q = d\lambda/dH$, λ is magnetostriction) and d are pseudo piezomagnetic and piezoelectric coefficients, respectively. Pseudo piezomagnetic coefficient is defined as the slope of magnetostriction versus magnetic field curve. For an ME composite, electric polarization due to the application of a magnetic field is found as:

$$\left(\frac{\partial \boldsymbol{P}}{\partial \boldsymbol{H}}\right)_{\boldsymbol{S}} = \boldsymbol{k_c q d} = \boldsymbol{\alpha} \tag{7}$$

where k_c is a mechanical coupling factor $(\mathbf{0}\ \ |\boldsymbol{k_c}|\ \ \mathbf{1})$ for the two phases and α is the ME susceptibility of the composite, from which we can calculate the ME voltage coefficient as:

$$\boldsymbol{\alpha_E} = \frac{\boldsymbol{\alpha}}{\boldsymbol{\varepsilon_o \varepsilon_r}} = \left(\frac{\partial \boldsymbol{E}}{\partial \boldsymbol{H}}\right)_{\boldsymbol{S}} \tag{8}$$

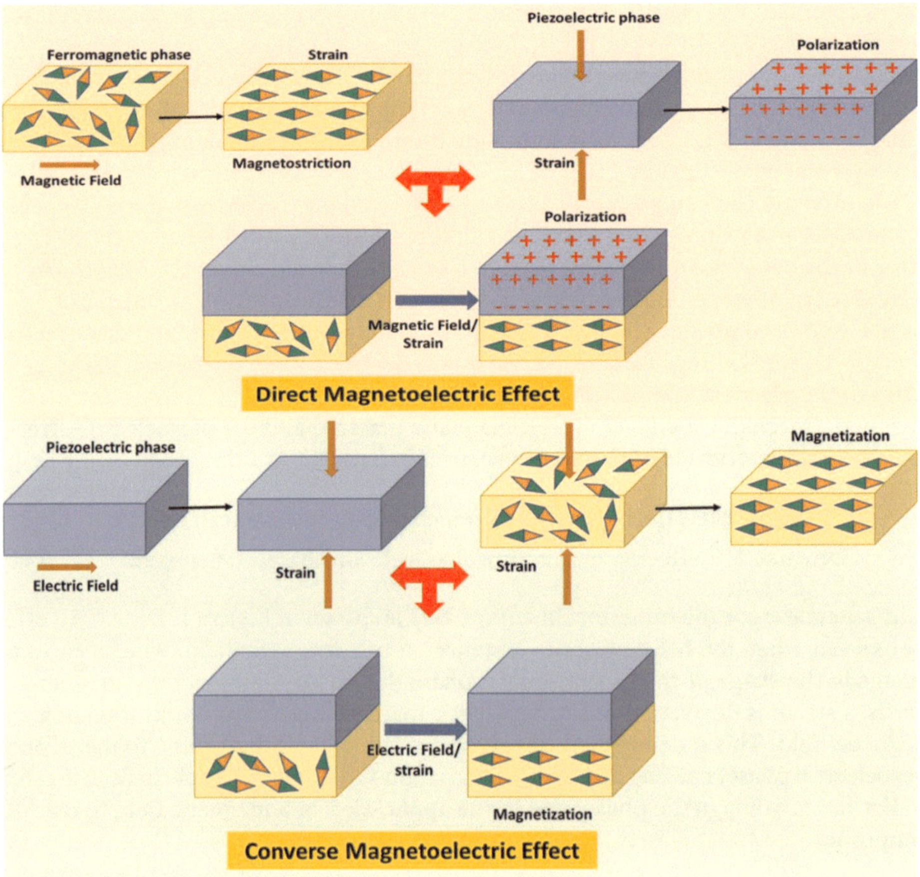

Figure 1.
Schematic diagram of strain-mediated ME effect in an ME composite.

where ε_o and ε_r are the electrical permittivity of free space and the dielectric constant of the piezoelectric phase, respectively.

Therefore, an entirely new property comes up in a composite of ferromagnetic and piezoelectric materials since neither of the constituent material shows a magnetoelectric effect. This product property-based ME response has resulted in due to the elastic coupling between the two constituent materials. This ME response of the composite, however, strongly depends on individual material characteristics such as the high pseudo piezomagnetic coefficient of ferromagnetic material and large piezoelectric coefficients of piezoelectric material along with strong mechanical coupling (large k_c) between the two materials. It can be said that the ME effect is an extrinsic property of these ME composites, which is the function of various extrinsic parameters such as the microstructure of the composite and how two phases couple magnetoelectrically at a ferromagnetic-piezoelectric interface. Moreover, this ME response in these composites is observed in ambient conditions, which makes these composites technically viable. Also, the ME response in these composites is several orders of magnitude larger when compared with single-phase magnetoelectric multiferroics. Various composites of ferromagnetic and piezoelectric materials have been investigated in recent years. These ME composites can become useful for practical applications by utilizing this extrinsic ME effect.

3. Candidate materials for ME composites

For an ME composite, the major requirements from a materials perspective are a high piezoelectric coefficient for the piezoelectric phase and large magnetostriction in the ferromagnetic phase. **Figure 2** shows piezoelectric coefficients and magnetostriction some of the constituent phases investigated for ME composites.

Many more candidate materials have been used to form ME composites. A ferromagnetic material can be chosen from a list of metallic FM materials, such Ni, Fe, or Co, from FM alloys, such as Terfenol-D, Metglas, and Permendure, or FM ceramics, such as Magnetic spinel ferrites, Garnets, and Manganites. All these materials have shown a good ME effect when combined with a piezoelectric material. A piezoelectric material is either an oxide ceramic, such as $BaTiO_3$, PZT, KNN, and PMN-PT, or a polymer such as polyvinylidene fluoride (PVDF).

3.1 Phase connectivity schemes

As mentioned earlier, strong ME response in ME composites above room temperature makes them technologically viable in various applications. Therefore, ME composites have been investigated using various combinations of materials and phase connectivity schemes, which include (1) bulk ceramic ME composites formed using piezoelectric ceramics and ferrites, (2) Bi-phase ME composites by combining FM alloys and piezoelectric materials, (3) three-phase (two FM phases and one piezoelectric or otherwise) ME composites, and (4) nanostructured thin films of ferroelectric and magnetic phases-based ME composite [26]. Newnham et al. introduced the concept of phase connectivity schemes (**Figure 3**) in which we use the notations 0–3, 2–2, and 1–3 for describing the structure of two phases of a bulk composite [27]. In these notations, each number represents the connectivity of the respective phase. For instance, 0–3 notation is used for a particulate composite in which particles of the FM phase (represented by 0) are engrained in the matrix of the piezoelectric phase (represented by 3). In 2–2 type composites, which we also call laminate composites, alternate layers (in form of thin sheets/films) of one phase are bonded with layers of another phase. In a 1–3 type composite, fibers/rods/wires/tubes of the magnetostrictive phase are embedded in the matrix of the piezoelectric phase. Each of these structures has its advantages and disadvantages.

Among these connectivity schemes, the most useful is the 2–2 type laminate composite due to its ease of fabrication and relatively better ME response. As most of

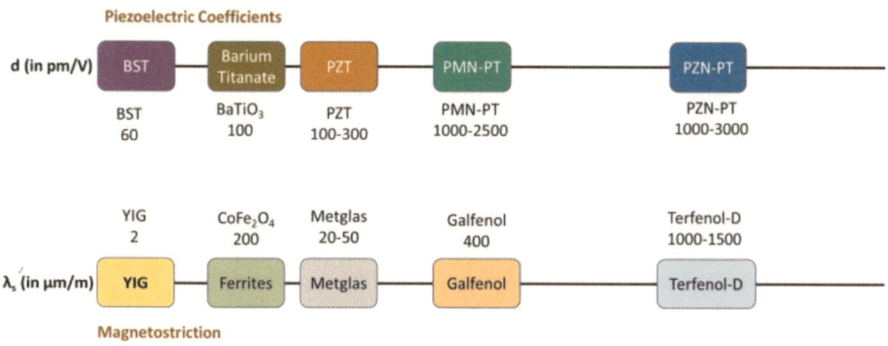

Figure 2.
Piezoelectric coefficients and magnetostriction of some of the constituent phases.

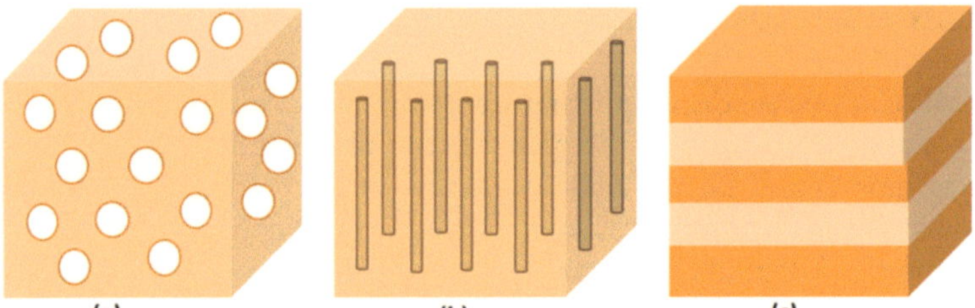

Figure 3.
Phase connectivity schemes of (a) 0–3 particulate bulk composite, (b) 1–3 fiber-matrix bulk composite, and (c) 2–2 laminate bulk composite.

the FM materials possess higher conductivity compared to piezoelectric materials (which are usually good dielectrics) in 0–3 type particulate composite, the inhomogeneous distribution of FM particles in piezoelectric matrix leads to more dielectric losses, which eventually reduces ME response. The issue of higher dielectric losses may be overcome in 1–3 type composites; however, the difficulty of fabrication makes these composites less viable. Moreover, their ME response is also relatively low compared to 2–2 type laminate composite. The connectivity schemes can also be employed to make ME composite thin films on suitable substrates. However, the ME response in these ME composite thin films is not as good as it is in their bulk counterpart. This may be attributed to the relatively lower piezoelectric effect in piezoelectric thin films.

4. Applications of ME composites

ME composites find numerous applications from energy harvesting to AC/DC magnetic field sensors, resonators, phase shifters, ME transformers, etc. Here, we will focus on the energy-harvesting aspect of these ME composites. As mentioned earlier, in an ME composite an electric field can be generated by the application of a magnetic field across the composite, which is termed the direct magnetoelectric effect. Now to produce this direct ME effect, both AC and DC magnetic fields need to be applied across the composite. DC magnetic field is the bias field that produces strain in the magnetostrictive phase through magnetostriction, while the AC field is required to lift the time-reversal symmetry of the magnetostrictive phase. The sweeping of the DC magnetic field initially increases the magnetostriction, which finally reaches saturation at a specific value of the magnetic field for a particular FM material. To generate continuous electrical output from piezoelectric material, we need to subject it to periodic stress. Therefore, in an ME composite, the application of an AC magnetic field leads to the production of periodic stresses in the piezoelectric phase, which is connected to the FM phase ***via*** strain-mediated magnetoelectric coupling. Thus, an ME composite needs to be subjected to a low-frequency AC magnetic field of small magnitude to generate continuous ME output. This low-frequency weak AC magnetic field is omnipresent in the proximity of power cables. A power cable carrying 50 A alternating current of the frequency 50–60 Hz will generate an alternating magnetic field of about 10 Oe at a distance of 1 cm [28]. This is a weak magnetic field that is usually considered noise and it is also detrimental to the human body. This unused and wasted magnetic field energy could be converted into electrical energy, which will be useful for powering low-power devices.

4.1 ME composites-based cantilever

There have been several designs proposed for effectively harvesting the electrical energy from the magnetic field and acoustic vibrations. The most useful design is to have an ME composite cantilever in which an FM layer is bonded with a piezoelectric layer as a 2–2 type laminate composite. In the case of a cantilever made from piezoelectric material only and subjected to vibrations, a tip mass is used to actuate it. In an ME composite cantilever, the tip mass is replaced with a permanent magnet. On application of an alternating magnetic field across this ME cantilever, the magnetic energy is transformed into vibrations and then into electricity through the piezoelectric effect. A schematic of such an ME cantilever along with a mechanism of the ME effect is shown in **Figure 4**.

Using the above design, Liu et al. [29] demonstrated a copper-based piezoelectric twin beam with a NdFeB permanent magnet of dimensions $30 \times 6 \times 0.3$ mm^3. A maximum output power density of 11.73 μW cm^{-3} Oe^{-1} was attained with a 100 Hz alternating magnetic field. A similar design was used by Lu et al. [30] for a composite beam of piezoelectric single crystal and Ni. They achieved a power density of 270 μW cm^{-3} Oe^{-1} in a 50 Hz AC magnetic field. This kind of design has attracted a lot of attention for energy harvesting applications due to its flexibility and ease of fabrication [31, 32].

4.2 Resonant condition ME composite cantilever

Further enhancement of conversion efficiency can be achieved using ME composites at resonant conditions where the frequency of the AC magnetic field is matched with the natural frequency of the ME composite cantilever. This is usually achieved by making light and slender cantilevers formed from a good mechanical property magnetostrictive material and a strong piezoelectric material. This offers the advantage that it is capable of not only converting vibration energy but also it can harness the magnetic field energy at a low frequency using the ME effect. A schematic of such a dual-phase energy harvester is shown in **Figure 5**. Zhou et al. [30] demonstrated a dual-phase energy harvester of Ni and piezoelectric macro-fiber composites, which produced a power density of 4.5 mW cm^{-3} G^{-1} at resonance.

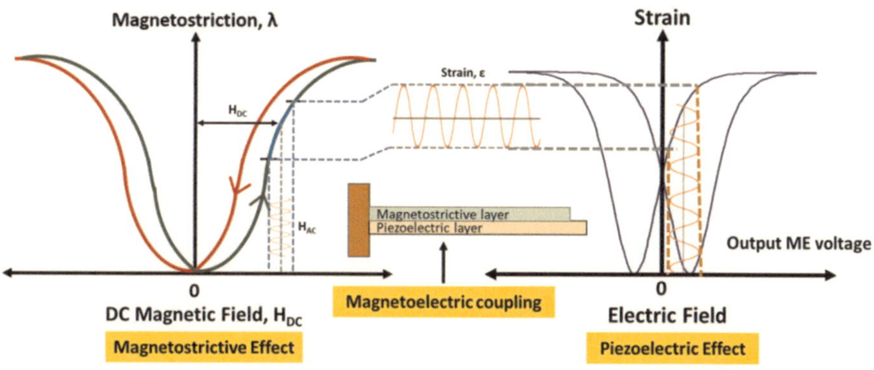

Figure 4.
Working principle of magnetoelectric cantilever-based energy harvester.

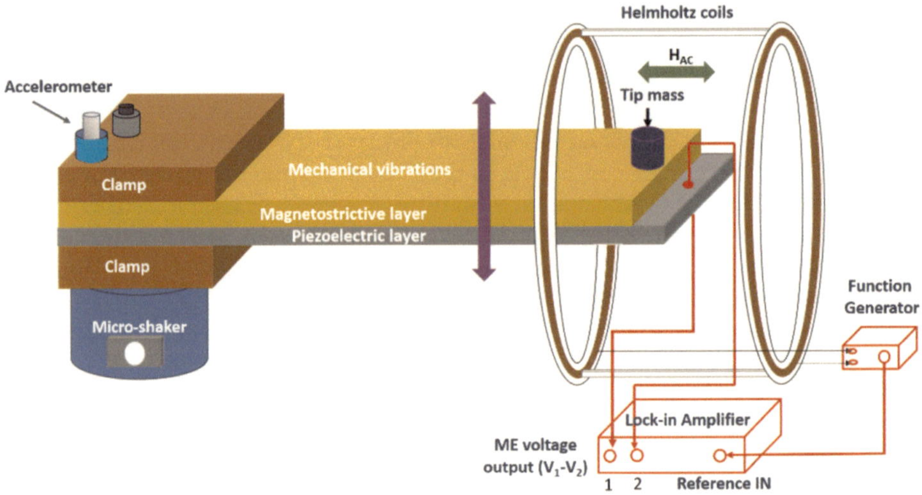

Figure 5.
A self-biased dual-phase ME energy harvester.

4.3 Magneto-mechano-electric (MME)-based ME composite energy harvester

As mentioned earlier, if the tip mass in an ME composite cantilever is replaced with a magnet, it is called a magneto-mechano-electric (MME) component. A schematic of an MME is shown in **Figure 6**.

In this MME component, electrical energy generation is ascribed to three mechanisms: magnetostriction in the FM layer due to the magnets and the deformation of the piezoelectric layer due to magnetostrictive strain, and AC magnetic field-induced vibrations. An MME component of nickel and piezoelectric single-crystal fibers which could generate a power density of 46 mW cm^{-3} Oe^{-1} in a magnetic field of 1.6 G was reported by Ryu et al. [33]. This MME component could ignite 35 LEDs. The output power density of ME composites mainly depends on constituent piezoelectric and magnetostrictive materials and also the extent of ME coupling between them. For example, Terfenol-D, which is known as the giant magnetostrictive material and has saturated magnetostriction of around 1500–2000 ppm under the driving magnetic

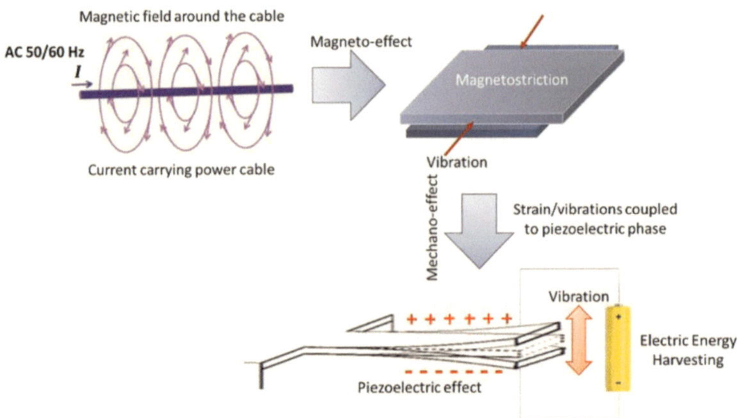

Figure 6.
Schematic of an MME generator.

field, was used in early ME composite studies [34]. However, its brittleness makes it inviable for low-frequency magnetic field environments due to its higher natural frequency. Moreover, Terfenol-D requires a DC magnetic field of 2–3 kOe to reach saturation magnetostriction [35]. A strong DC magnetic field often leads to bulky structures and usually causes electromagnetic interference as well. Therefore, ME composites requiring a small bias field caught attention [36]. Chu et al. [37] reported Metglas and piezoelectric single-crystal fibers ME composites, which were longitudinally magnetized and poled (L-T) in the transverse direction. A large ME coefficient of 22.92 V cm^{-1} Oe^{-1} for a 6 Oe DC bias field could be obtained. However, achieving resonance at low frequency is rather difficult. Therefore, it may be concluded that all these combinations come with certain advantages and disadvantages [38, 39].

5. Conclusions

The ME composites of a ferromagnetic material and a piezoelectric material use the magnetoelectric effect in conjunction with the piezoelectric effect in electrical energy harvesting. They have shown better conversion efficiency as compared to individual piezoelectric energy harvesters. However, the choice of constituent materials and their connectivity schemes play a crucial role in designing these ME composite energy harvesters. Further, enhancement in energy harvesting properties in these ME composites may be achieved by using different phase connectivity schemes. Although the 1–3 phase connectivity scheme has shown significant improvement in the ME effect, the difficulty in the fabrication of these 1–3 type composites is still a hindrance.

Acknowledgements

The authors would like to thank VIT management for their encouragement and support.

Conflict of interest

Authors have no mutual conflict of interest.

Appendices and nomenclature

FM	Ferromagnetic
FE	Ferroelectric
PE	Piezoelectric
ME	Magnetoelectric
MME	Magneto-mechano-electric
BST	Barium strontium titanate
YIG	Yttrium iron garnet
PZT	Lead zirconium titanate
PMN-PT	Lead magnesium niobate-lead titanate
PZN-PT	Lead zinc niobate-lead titanate

Author details

Tarun Garg* and Lickmichand M. Goyal
Department of Physics, School of Advanced Sciences, Vellore Institute of Technology, Vellore, India

*Address all correspondence to: gargphy1981@gmail.com

References

[1] Wang Y, Atulasimha J. Nonlinear magnetoelectric model for laminate piezoelectric–magnetostrictive cantilever structures. Smart Materials and Structures. 2012;**21**:085023. DOI: 10.1088/0964-1726/21/8/085023

[2] Jafari H, Ghodsi A, Azizi S, Ghazavi MR. Energy harvesting based on magnetostriction, for low frequency excitations. Energy. 2017;**124**:1-8. DOI: 10.1016/j.energy.2017.02.014

[3] Glynne-Jones P, Tudor MJ, Beeby SP, White NM. An electromagnetic, vibration-powered generator for intelligent sensor systems. Sensors and Actuators A: Physical. 2004;**110**: 344-349. DOI: 10.1016/j. sna.2003.09.045

[4] Galchev T, Kim H, Najafi K. Micro power generator for harvesting low-frequency and nonperiodic vibrations. Journal of Microelectromechanical Systems. 2011;**20**(4):852-866. DOI: 10.1109/JMEMS.2011.2160045

[5] Yang B, Lee C, Xiang W, Xie J, Han He J, Kotlanka RK, et al. Electromagnetic energy harvesting from vibrations of multiple frequencies. Journal of Micromechanics and Microengineering. 2009;**19**:035001. DOI: 10.1088/ 0960-1317/19/3/035001

[6] Li W, Lu L, Kottapalli AGP, Pei Y. Bioinspired sweat-resistant wearable triboelectric nanogenerator for movement monitoring during exercise. Nano Energy. 2022;**95**:107018. DOI: 10.1016/j.nanoen.2022.107018

[7] Long L, Liu W, Wang Z, He W, Li G, Tang Q, et al. High performance floating self-excited sliding triboelectric nanogenerator for micro mechanical energy harvesting. Nature Communications. 2021;**12**:4689. DOI: 10.1038/s41467-021-25047-y

[8] Yin P, Aw KC, Jiang X, Xin C, Guo H, Tang L, et al. Fish gills inspired parallel-cell triboelectric nanogenerator. Nano Energy. 2022;**95**:106976. DOI: 10.1016/j. nanoen.2022.106976

[9] Wang H, Hu F, Wang K, Liu Y, Zhao W. Three-dimensional piezoelectric energy harvester with spring and magnetic coupling. Applied Physics Letters. 2017;**110**:163905. DOI: 10.1063/1.4981885

[10] Roundy S, Wright PK, Rabaey J. A study of low level vibrations as a power source for wireless sensor nodes. Computer Communications. 2003;**26**:1131-1144. DOI: 10.1016/S0140-3664(02)00248-7

[11] Zhou S, Cao J, Inman DJ, Lin J, Liu S, Wang Z. Broadband tristable energy harvester: Modeling and experiment verification. Applied Energy. 2014;**133**: 33-39. DOI: 10.1016/j. apenergy.2014.07.077

[12] Song J, Wang DF, Shan G, Ono T, Itoh T, Shimoyama I, et al. A black gauze cap-shaped bistable energy harvester with a movable design for broadening frequency bandwidth. Smart Materials and Structures. 2020;**29**:025015. DOI: 10.1088/1361-665X/ab6077

[13] Lu Z-Q, Chen J, Ding H, Chen L-Q. Two-span piezoelectric beam energy harvesting. International Journal of Mechanical Sciences. 2020;**175**:105532. DOI: 10.1016/j.ijmecsci.2020.105532

[14] Lu Z-Q, Zhang F-Y, Fu H-L, Ding H, Chen L-Q. Rotational nonlinear double-beam energy harvesting. Smart Materials and Structures. 2022;**31**:025020. DOI: 10.1088/1361-665X/ac4579

[15] Lu C, Jiang X, Li L, Zhou H, Yang A, Xin M, et al. Wind energy harvester using piezoelectric materials. Review of Scientific Instruments. 2022;**93**:031502. DOI: 10.1063/5.0065462

[16] Liu F-R, Zhang W-M, Zhao L-C, Zou H-X, Tan T, Peng Z-K, et al. Performance enhancement of wind energy harvester utilizing wake flow induced by double upstream flat-plates. Applied Energy. 2020;**257**:114034. DOI: 10.1016/j.apenergy.2019.114034

[17] Lu Z-Q, Chen J, Ding H, Chen L-Q. Energy harvesting of a fluid-conveying piezoelectric pipe. Applied Mathematical Modelling. 2022;**107**:165-181. DOI: 10.1016/j.apm.2022.02.027

[18] Zou H-X, Li M, Zhao L-C, Gao Q-H, Wei K-X, Zuo L, et al. A magnetically coupled bistable piezoelectric harvester for underwater energy harvesting. Energy. 2021;**217**:119429. DOI: 10.1016/j.energy.2020.119429

[19] Pradhan DK, Kumari S, Rack PD, Kumar A. Applications of strain-coupled Magnetoelectric composites. In: Encyclopedia of Smart Materials. Amsterdam, Netherlands: Elsevier; 2022. pp. 229-238. DOI: 10.1016/B978-0-12-815732-9.00048-6

[20] Landau LD, Lifshitz EM. Chapter IV - static magnetic field. In: Landau LD, Lifshitz EM, (Eds). Electrodynamics of Continuous Media. (Second Edition). vol. 8. Amsterdam: Pergamon; 1984. pp. 105-129. DOI: 10.1016/B978-0-08-030275-1.50010-2

[21] Rivera J-P. On definitions, units, measurements, tensor forms of the linear magnetoelectric effect and on a new dynamic method applied to Cr-Cl boracite. Ferroelectrics. 1994;**161**:165-180. DOI: 10.1080/00150199408213365

[22] Curie P. On the symmetry in the physical phenomena. Journal de Physique. 1894;**3**:393-415

[23] Astrov DN. Magnetoelectric effect in chromium oxide. Soviet Physics JETP. 1961;**13**:729-733

[24] van Suchtelen J. Product properties: A new application of composite materials. Philips Research Reports. 1970;**27**:28-37

[25] Nan C-W. Magnetoelectric effect in composites of piezoelectric and piezomagnetic phases. Physical Review B. 1994;**50**:6082-6088. DOI: 10.1103/PhysRevB.50.6082

[26] Nan C-W, Bichurin MI, Dong S, Viehland D, Srinivasan G. Multiferroic magnetoelectric composites: Historical perspective, status, and future directions. Journal of Applied Physics. 2008;**103**:031101. DOI: 10.1063/1.2836410

[27] Newnham RE, Bowen LJ, Klicker KA, Cross LE. Composite piezoelectric transducers. Materials and Design. 1980;**2**:93-106. DOI: 10.1016/0261-3069(80)90019-9

[28] Halgamuge MN, Abeyrathne CD, Mendis P. Measurement and analysis of electromagnetic fields from trams, trains and hybrid cars. Radiation Protection Dosimetry. 2010;**141**:255-268. DOI: 10.1093/rpd/ncq168

[29] Liu G, Ci P, Dong S. Energy harvesting from ambient low-frequency magnetic field using magneto-mechano-electric composite cantilever. Applied Physics Letters. 2014;**104**:032908. DOI: 10.1063/1.4862876

[30] Lu Y, Chen J, Cheng Z, Zhang S. The PZT/Ni unimorph magnetoelectric energy harvester for wireless sensing

applications. Energy Conversion and Management. 2019;**200**:112084. DOI: 10.1016/j.enconman.2019.112084

[31] Annapureddy V, Palneedi H, Hwang G-T, Peddigari M, Jeong D-Y, Yoon W-H, et al. Magnetic energy harvesting with magnetoelectrics: An emerging technology for self-powered autonomous systems. Sustain Energy Fuels. 2017;**1**:2039-2052. DOI: 10.1039/C7SE00403F

[32] Costa P, Nunes-Pereira J, Pereira N, Castro N, Gonçalves S, Lanceros-Mendez S. Recent Progress on piezoelectric, pyroelectric, and Magnetoelectric polymer-based energy-harvesting devices. Energy Technology. 2019;7:1800852. DOI: 10.1002/ente.201800852

[33] Ryu J, Kang JE, Zhou Y, Choi SY, Yoon WH, Park DS, et al. Ubiquitous magneto-mechano-electric generator. Energy Environmental Sciences. 2015;**8**: 2402-2408. DOI: 10.1039/c5ee00414d

[34] Yang J, Yue X, Wen Y, Li P, Yu Q, Bai X. Design and analysis of a 2D broadband vibration energy harvester for wireless sensors. Sensors and Actuators A: Physical. 2014;**205**:47-52. DOI: 10.1016/j.sna.2013.10.005

[35] Ryu J, Priya S, Carazo AV, Uchino K, Kim H-E. Effect of the Magnetostrictive layer on Magnetoelectric properties in Lead Zirconate Titanate/Terfenol-D laminate composites. Journal of the American Ceramic Society. 2001;**84**: 2905-2908. DOI: 10.1111/j.1151-2916.2001.tb01113.x

[36] Kambale RC, Yoon W-H, Park D-S, Choi J-J, Ahn C-W, Kim J-W, et al. Magnetoelectric properties and magnetomechanical energy harvesting from stray vibration and electromagnetic wave by $Pb(Mg_{1/3}Nb_{2/3})O_3$-$Pb(Zr, Ti)O_3$ single crystal/Ni cantilever. Journal of Applied Physics. 2013;**113**:204108. DOI: 10.1063/1.4804959

[37] Chu Z, Shi H, Shi W, Liu G, Wu J, Yang J, et al. Enhanced resonance Magnetoelectric coupling in (1-1) connectivity composites. Advanced Materials. 2017;**29**:1606022. DOI: 10.1002/adma.201606022

[38] Wang Y, Hu J, Lin Y, Nan C-W. Multiferroic magnetoelectric composite nanostructures. NPG Asia Materials. 2010;**2**:61-68. DOI: 10.1038/asiamat.2010.32

[39] Ren Z, Tang L, Zhao J, Zhang S, Liu C, Zhao H. Comparative study of energy harvesting performance of magnetoelectric composite-based piezoelectric beams subject to varying magnetic field. Smart Materials and Structures. 2022;**31**(10):105001. DOI: 10.1088/1361-665X/ac798c

Chapter 5

Equivalent Circuit Model of Magnetostrictive/Piezoelectric Laminated Composite

María Sol Ruiz and Adrián César Razzitte

Abstract

An equivalent circuital model of magnetostrictive/piezoelectric laminated composite has been developed in order to predict its behavior in presence of dynamic electromagnetic fields. From magnetostrictive and piezoelectric constitutive equations, and using an equation of motion, magnetic-mechanical-electric equations are: obtained by building a symmetric adhoc equivalent circuit about the magnetoelectric (ME) coupling. The coefficients of the direct and converse effects are simulated. The circuit is further used to predict the voltage coefficients of laminated composite. The multilayer material is found to have significantly higher ME coefficients near resonance frequency. The ME coefficients and the voltage coefficients change significantly with the configuration of the multilayer, more specifically when the laminate operates in longitudinal-transverse (L-T) and transverse-transverse (T–T) modes.

Keywords: magnetoelectric effect, magnetoelectric voltage coefficient, magnetostrictive, piezoelectric, laminated composite

1. Introduction

The Magnetoelectric (ME) effect arises in multiferroic materials that are electrically and magnetically polarizable due to coupling between electrical polarization and magnetization. These materials are interesting for technological applications, such as magnetic field sensors, transducers, resonators, and devices that provide opportunities in the area of renewable and sustainable energy through energy harvesting [1, 2].

Although the coupling can have nonlinear components, the ME effect is usually described mathematically by the linear ME coupling coefficient, which is the dominant coupling term [3]. The ME effect is characterized by an electric polarization induced by an applied magnetic field or a magnetization induced by an applied electric field called direct magnetoelectric effect (DME) and converse magnetoelectric effect (CME), respectively. The coefficients of the direct (α_{DME}) and converse (α_{CME}) effects can be expressed as:

$$\alpha_{DME} = \frac{\partial P}{\partial H} \quad (1)$$

$$\alpha_{CME} = \frac{\partial B}{\partial E} \tag{2}$$

being E electric field intensity, B magnetic flux density, P electric polarization, and H magnetic field intensity.

Maxwell's equations of classical electromagnetism state that, in free space, E and B are not independent but are intimately linked to each other[1]. Similarly, a theoretical method for the solid state that simultaneously describes relationships $P \sim H$ and $B \sim E$ (i.e. it describes the DME and CME behaviors) is developed below.

Multilayered composites can have even higher magnetoelectric coefficients than the particulate composites [4] and are therefore considered as viable alternatives [5].

2. Modes of operation and vibrational modes

The system to be modeled consists of bonded ferrite-ferroelectric layers of rectangular geometry. The dimensions are considered to be such that the width w and thickness t are very small relative to the length ℓ; that is, the system is considered as one-dimensional (1D). The thicknesses of the magnetostrictive and piezoelectric layers are designated as t_m and t_p, respectively. It is assumed that the interface (boundary between layers) is continuous.

In the modeling of ME laminate, different modes of operation or configurations need to be considered; in addition, to properly defining the coordinate systems (local and global) to be used. For historical reasons, the modes of operation find different lexicons. To prevent confusion, the notation adopted in this work is detailed below. Four basic modes of operation are distinguished and defined, according to the directions of sensed[2] magnetic field (H) and electric field (E). When the magnetic and electric field directions coincide with the longitudinal direction of the laminate, the mode is defined as "L-L". Similarly, the modes "L-T," "T-L," and "T–T" designate longitudinal magnetic field - transverse electric field, transverse magnetic field - longitudinal electric field, and transverse magnetic field - transverse electric field, respectively. The basic modes of operation are outlined in a summary in **Figure 1**.

When a magnetic field is applied to an ME laminate, more than one magnetostrictive vibrational mode is excited. However, the problem can be simplified, without losing correlation with reality, by considering a material with dimensions such that its width and thickness are very small compared to its length. In such case, the longitudinal axis direction of the laminate can be considered as the reference direction. Thus, in this chapter, the main direction of vibration is the longitudinal direction of the laminate[3] (this justifies the selection of a one-dimensional 1D model).

Once the direction of vibration of the laminate is designated, it is then possible to carry out a consistent analysis of the system. When the material is excited with a

---

[1] The use of the existing connection between electricity and magnetism is what has generated the enormous impact known in the field of technology.

[2] It refers to "sensed" H since the lexicon is adopted from the field of application of magnetic sensors; in this notation, the first letter refers to the direction of the magnetic field and the second letter refers to the direction of the electric field.

[3] It is chosen to assign the main vibration direction of the laminate as the reference direction, following the definitions for piezoelectric vibration modes [6].

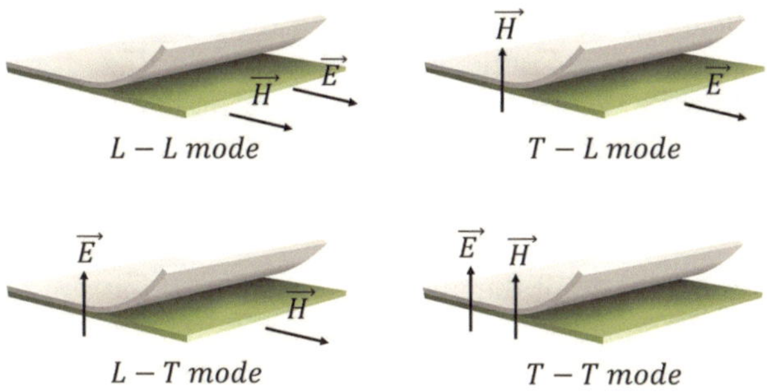

Figure 1.
Modes of operation of laminated magnetoelectric material.

magnetic field perpendicular to the longitudinal direction of the laminate, the fundamental mode of vibration will be the transverse mode and when it is excited with a magnetic field parallel to the longitudinal direction of the laminate, the fundamental mode of vibration will be the longitudinal mode (in both cases, as was mentioned, other modes of vibration can be considered as negligible).

3. Constitutive equations of the piezoelectric and magnetostrictive phases

If the system is considered in T–T mode of operation, the direction of polarization in the piezoelectric layer and the direction of magnetic field are along the thickness of the laminate (direction that is made to coincide with the cartesian coordinate axis z). The displacement $u = u(x)$ coincides with the longitudinal direction of the laminate (direction of the coordinate axis x). **Figure 2a** shows the directions of vectors D (electric displacement or electric flux density), B (magnetic flux density), S (mechanical strain), T (mechanical stress), along with the global coordinates x,y,z.

The ME effect of composite materials is due to the combined action of multiple intrinsic characteristics of the materials involved. In materials composed of two phases, two sets of constitutive equations are: required to describe the ME product property. Constitutive equations, in their most general form of the tensor type, are reduced to linear equations (both piezoelectric and piezomagnetic). The equations are: established from an electro/magneto-static point of view (i.e.,, for small excitation signals), and the simplifications are mainly a result of the crystal symmetry of the materials [7]. In the treatment of piezoelectric and piezomagnetic constitutive equations, the subscripts were established taking into account the reference axes used to define piezoelectric coefficients. To comply with this definition, the local coordinates

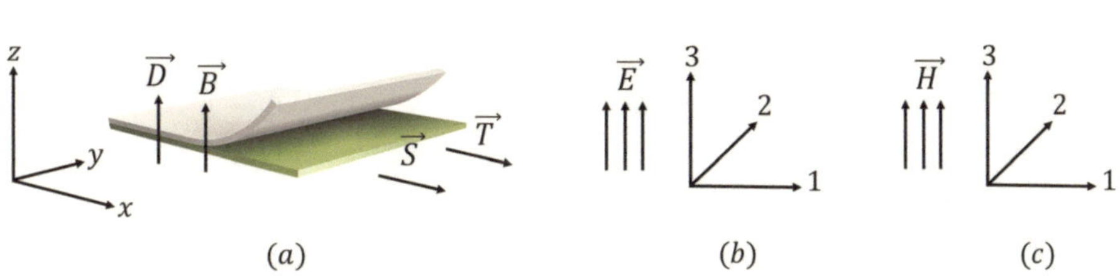

Figure 2.
(a) Schematic illustration of the laminated bilayer ME in T–T mode and local coordinates for the (b) piezoelectric and (c) piezomagnetic layer.

shown in **Figure 2b** are defined in the piezoelectric layer and the local coordinates shown in **Figure 2c** are defined in the piezomagnetic layer (i.e., for the magnetic material has established a direction of magnetization[4] in the direction of the axis 3).

3.1 Constitutive equations of the piezoelectric laminate

The constitutive equations of the piezoelectric material for the strain S_{1p} and the electric displacement D_3, are:

$$S_{1p} = s_{11}^{E} T_{1p} + d_{31,p} E_3 \quad (3)$$

$$D_3 = d_{31,p} T_{1p} + \varepsilon_{33}^{T} E_3 \quad (4)$$

Being E_3 the electric field in the piezoelectric layer along the z direction, T_{1p} the stress in the piezoelectric layer along the x direction, s_{11}^{E} the elastic compliance at constant E, $d_{31,p}$ the piezoelectric constant, and ε_{33}^{T} the dielectric permittivity at constant stress. The subscript p refers to the piezoelectric phase.

3.2 Constitutive equations of the magnetostrictive laminate

The constitutive equations of the magnetostrictive material for the strain S_{1m} and the magnetic flux density B_3, are:

$$S_{1m} = s_{11}^{H} T_{1m} + d_{31,m} H_3 \quad (5)$$

$$B_3 = d_{31,m} T_{1m} + \mu_{33}^{T} H_3 \quad (6)$$

Being H_3 the magnetic field intensity in the magnetostrictive layer along the z-direction, T_{1m} the stress in the piezomagnetic layer along the x direction, s_{11}^{H} the elastic compliance to constant H, $d_{31,m}$ the piezomagnetic constant, and μ_{33}^{T} the magnetic permeability at constant stress. The subscript m refers to the magnetostrictive phase.

There is a widely used notation in the design of laminated sensors in which the fundamental vibration mode is designated by two numbers: the first number indicates the direction of the excitation field and the second number indicates the direction of the longitudinal axis (both defined through local coordinates). Taking into account the **Figure 2**, the fundamental vibration mode of the magnetostrictive phase can be called "31-mode" and the fundamental vibration mode of the piezoelectric phase can be called "31-mode". That is, the phases of the composite material will be modeled as magnetostrictive and piezoelectric transducers, both in mode 31. Note the correspondence that exists (for each phase) between its fundamental mode of vibration and the subscripts of its "piezo-" parameters in the constitutive equations[5].

---

[4] Although strictly speaking, magnetization in ferrites is a magnetic polarization phenomenon, and polarization in piezoelectrics is an electrical polarization phenomenon, here "magnetization" is used for the magnetostrictive layer and "polarization" for the piezoelectric layer.

[5] The term "piezo-" refers to the piezoelectric or piezomagnetic coefficients, which are precisely electro-mechanical and magneto-mechanical parameters, respectively.

4. Equation of motion: general solution and boundary conditions

In order to design laminates operating under dynamic conditions, the coupling between the composite layers is considered through an equation of motion. Considering that the applied external field (electric or magnetic) is sinusoidal, the vibrational movement of the laminate will also be sinusoidal (harmonic oscillator).

The differential equation of motion for any element of mass can be obtained using Newton's second law, written as:

$$\overline{\rho}\frac{\partial^2 u(x,t)}{\partial t^2} = r_m\frac{\partial T_{1m}}{\partial x} + r_p\frac{\partial T_{1p}}{\partial x} \tag{7}$$

Where $u(x,t)$ is the displacement function of the medium and r_m and r_p are the relative relations of thickness of the magnetostrictive and piezoelectric layers, respectively, that is: $r_m = t_m/t$ and $r_p = t_p/t$, and it is evident that $r_m + r_p = 1$. Furthermore, $\overline{\rho} = r_p\,\rho_p + r_m\,\rho_m$ is the average mass density, with ρ_p and ρ_m as the mass densities of the piezoelectric and magnetostrictive components, respectively.

In the 1D model, mechanical movements are considered only in the longitudinal direction of the laminate (x-axis). In the direction perpendicular to it, the component layers are considered as free bodies and for the model, there are no strains or stresses between the layers of the composite.

The magnetostrictive and piezoelectric mass elements share the same displacement component u and the same strain component, that is:

$$u_m = u_p = u \tag{8}$$

$$S_{1m} = S_{1p} = S = \frac{\partial u}{\partial x} \tag{9}$$

From Eq. (3) and (5), the mechanical stresses in the phases can be expressed as:

$$T_{1p} = \frac{S_{1p}}{s_{11}^E} - \frac{d_{31,p}}{s_{11}^E}E_3 \tag{10}$$

$$T_{1m} = \frac{1}{s_{11}^H}S_{1m} - \frac{d_{31,m}}{s_{11}^H}H_3 \tag{11}$$

Substituting in Eq. (7) the derivatives of Eq. (10) and (11) and considering that $(\partial E_3)/\partial x = (\partial H_3)/\partial x = 0$ (since both E_3 and H_3 do not vary along the length of the material), the equation of motion is written as:

$$\frac{\partial^2 u(x,t)}{\partial t^2} = \overline{v}^2\frac{\partial^2 u(x,t)}{\partial x^2} \tag{12}$$

where $\overline{v}$ is the speed of sound in the composite material:

$$\overline{v}^2 = \left(\frac{r_m}{s_{11}^H} + \frac{r_p}{s_{11}^E}\right)/\overline{\rho} \tag{13}$$

Eq. (12) is the classical wave equation of motion. Your resolution makes it simplified to:

$$\frac{\partial^2 u(x)}{\partial x^2} + k^2 u(x) = 0 \tag{14}$$

Which has the form of the equation of motion of a simple harmonic oscillator but with the spatial variable instead of temporal. In addition, the so-called dispersion relation $k = \omega/\bar{v}$ holds, where k is the wave number and ω is the angular frequency.

The general solution of Eq. (14) is:

$$u(x) = A\cos kx + B\sin kx \tag{15}$$

To solve the integration constants A and B, the boundary conditions expressed in terms of displacement speeds $\dot{u}$ are established: considering that the ends of the material are free of external stresses, let $\dot{u}_1$ be the displacement speed at $x = 0$ and $\dot{u}_2$ the displacement speed in $x = \ell$, then:

$$\dot{u}_1 = j\omega u(0) \tag{16}$$

$$\dot{u}_2 = j\omega u(\ell) \tag{17}$$

The resolution leads to the expression:

$$u(x) = \frac{\dot{u}_1}{j\omega}\cos kx + \frac{\dot{u}_2 - \dot{u}_1 \cos k\ell}{j\omega \sin k\ell}\sin kx \tag{18}$$

And, according to Eq. (9) and (18), the strains at the faces of the laminate are:

$$S(x=0) = \frac{\dot{u}_2 - \dot{u}_1 \cos k\ell}{j\bar{v}\sin k\ell} \tag{19}$$

$$S(x=\ell) = \frac{\dot{u}_2 \cos k\ell - \dot{u}_1}{j\bar{v}\sin k\ell} \tag{20}$$

5. Potential: current coupling

Substituting Eq. (10) in Eq. (4), is obtained:

$$D_3 = d_{31,p}\left(\frac{S_{1p}}{s_{11}^E} - \frac{d_{31,p}}{s_{11}^E}E_3\right) + \varepsilon_{33}^T E_3 \tag{21}$$

By definition of electric potential and taking into account the geometry of the system, it can be expressed that the piezoelectric vibrator is satisfied: $V_p = t_p\, E_3$. Furthermore, if defined:

$$\overline{\varepsilon_{33}} = \varepsilon_{33}^T\left(1 - \frac{d^2{}_{31,p}}{s_{11}^E \varepsilon_{33}^T}\right) \tag{22}$$

Then the electrical displacement can be written as:

$$D_3 = \frac{d_{31,p}}{s_{11}^E} S_{1p} + \overline{\varepsilon_{33}} E_3 \tag{23}$$

The electric charge at the surface of the piezoelectric is obtained integrating over the surface, according to:

$$Q = \int_0^{\ell}\int_0^{w} D_3\, dxdy = w\frac{d_{31,p}}{s_{11}^E}[u(\ell) - u(0)] + w\ell\overline{\varepsilon_{33}}E_3 \tag{24}$$

From Eq. (18), the expressions for $u(\ell)$ and $u(0)$ are obtained according to:

$$u(\ell) = \frac{\dot{u}_1}{j\omega}\cos k\ell + \frac{\dot{u}_2 - \dot{u}_1\cos k\ell}{j\omega\sin k\ell}\sin k\ell = \frac{\dot{u}_2}{j\omega} \tag{25}$$

$$u(0) = \frac{\dot{u}_1}{j\omega}\cos k0 + \frac{\dot{u}_2 - \dot{u}_1\cos k\ell}{j\omega\sin k\ell}\sin k0 = \frac{\dot{u}_1}{j\omega} \tag{26}$$

Then:

$$Q = \int_0^{\ell}\int_0^{w} D_3\, dxdy = w\frac{d_{31,p}}{j\omega s_{11}^E}(\dot{u}_2 - \dot{u}_1) + w\ell\overline{\varepsilon_{33}}E_3 \tag{27}$$

Taking into account the properties of harmonic phasor (the analysis is in the field of angular frequency, not temporal), the electric current in the piezoelectric vibrator I_p can be expressed as:

$$I_p = \frac{dQ}{dt} = j\omega Q \tag{28}$$

Defining the static capacitance in the piezoelectric layer C_0 and the electromechanical coupling factor φ_p as:

$$C_0 = \frac{\overline{\varepsilon_{33}}w\ell}{t_p} \tag{29}$$

$$\varphi_p = w\frac{d_{31,p}}{s_{11}^E} \tag{30}$$

And substituting Eq. (29) and (30) in Eq. (28), the expression of coupling for piezoelectric material is:

$$I_p = j\omega C_0 V_p + \varphi_p(\dot{u}_2 - \dot{u}_1) \tag{31}$$

Similarly, from the constitutive equations of the magnetostrictive material, replacing Eq. (11) in (6), is obtained:

$$B_3 = \left(\frac{d_{31,m}}{s_{11}^H}\right)S_{1m} + \overline{\mu_{33}^T}H_3 \tag{32}$$

where is defined:

$$\overline{\mu_{33}^T} = \mu_{33}^T\left(1 - \frac{d^2{}_{31,m}}{\mu_{33}^T s_{11}^H}\right) \tag{33}$$

According to Faraday's law of induction, on the solenoid V_m can be expressed as:

$$V_m = -N\frac{d\Phi}{dt} \tag{34}$$

where $\Phi = B_3 w t_m$ is the magnetic flux. And under concepts of harmonic phasors, the differential of the magnetic flux can be expressed as:

$$V_m = -Nj\omega\Phi = -Nj\omega B_3 w t_m \tag{35}$$

Substituting Eq. (32) in Eq. (35):

$$V_m = -Nj\omega w t_m\left[\left(\frac{d_{31,m}}{s_{11}^H}\right)S_{1m} + \overline{\mu_{33}^T}H_3\right] \tag{36}$$

Considering a solenoid with N turns of total length ℓ and through which an electric current I_m circulates, the generated magnetic field in the solenoid can be expressed as:

$$H_3 = \frac{N}{\ell}I_m \tag{37}$$

And taking into account that the relationship between deformation and displacement velocities can be expressed by the relation [8]:

$$S_{1m} = \frac{\dot{u}_2 - \dot{u}_1}{j\omega\ell} \tag{38}$$

Then:

$$V_m = -\frac{N w t_m}{\ell}\frac{d_{31,m}}{s_{11}^H}(\dot{u}_2 - \dot{u}_1) - j\omega\frac{w t_m N^2}{\ell}\overline{\mu_{33}^T}I_m \tag{39}$$

Defining the static inductance in the magnetostrictive layer L_0 and the magneto-mechanical coupling factor φ_m as:

$$L_0 = \overline{\mu_{33}^T}N^2\frac{w t_m}{\ell} \tag{40}$$

$$\varphi_m = w t_m\frac{N}{\ell}\frac{d_{31,m}}{s_{11}^H} \tag{41}$$

The expression of coupling for magnetostrictive material is:

$$V_m = -\varphi_m(\dot{u}_2 - \dot{u}_1) - j\omega L_0 I_m \tag{42}$$

6. Magneto-mechano-electric coupling

Based on the constitutive equations of both components (Section 3) and the equation of motion (via strain–stress coupling between layers; Section 4), an equivalent circuit model for laminated ME is developed. The mechanical forces are relative to the stresses. If the forces at the ends of the laminate are called F_1 at $x = 0$ and F_2 at $x = \ell$, it is satisfied that:

$$-F_1 = wt_p \, T_{1p}\big|_{x=0} + wt_m \, T_{1m}|_{x=0} \tag{43}$$

$$-F_2 = wt_p \, T_{1p}\big|_{x=\ell} + wt_m \, T_{1m}|_{x=\ell} \tag{44}$$

Substituting Eq. (10), (11), (19), and (20) in Eq. (43) and (44) and taking into account Eq. (18), the expressions are: obtained:

$$F_1 = -\left(\frac{wt_p}{s_{11}^E} + \frac{wt_m}{s_{11}^H}\right)\left(\frac{\dot{u}_2 - \dot{u}_1 \cos k\ell}{j\overline{v}\sin k\ell}\right) + wt_p \frac{d_{31,p}}{s_{11}^E} E_3 + wt_m \frac{d_{31,m}}{s_{11}^H} H_3 \tag{45}$$

$$F_2 = -\left(\frac{wt_p}{s_{11}^E} + \frac{wt_m}{s_{11}^H}\right)\left(\frac{\dot{u}_2 \cos k\ell - \dot{u}_1}{j\overline{v}\sin k\ell}\right) + wt_p \frac{d_{31,p}}{s_{11}^E} E_3 + wt_m \frac{d_{31,m}}{s_{11}^H} H_3 \tag{46}$$

Furthermore, with the equations obtained in Section 5, the force expressions can be simplified to:

$$F_1 = -\overline{\rho}\overline{v}wt\left(\frac{\dot{u}_2 - \dot{u}_1 \cos k\ell}{j\sin k\ell}\right) + \varphi_p V_p + \varphi_m I_m \tag{47}$$

$$F_2 = -\overline{\rho}\overline{v}wt\left(\frac{\dot{u}_2 \cos k\ell - \dot{u}_1}{j\sin k\ell}\right) + \varphi_p V_p + \varphi_m I_m \tag{48}$$

If mechanical impedances are defined as:

$$Z_1 = \frac{\overline{\rho}\,\overline{v}\,wt}{j\sin k\ell} \tag{49}$$

$$Z_2 = j\overline{\rho}\,\overline{v}\,wt\tan\frac{k\ell}{2} \tag{50}$$

Then the mechanical forces are rewritten as:

$$F_1 = Z_1(\dot{u}_1 - \dot{u}_2) + Z_2\dot{u}_1 + \varphi_p V_p + \varphi_m I_m \tag{51}$$

$$F_2 = Z_1(\dot{u}_1 - \dot{u}_2) - Z_2\dot{u}_2 + \varphi_p V_p + \varphi_m I_m \tag{52}$$

The ME equivalent circuit corresponding to the magneto-mechano-electric expressions obtained is represented in **Figure 3**. In this circuit, the forces F_1 and F_2 act as "mechanical voltage" and the displacements $\dot{u}_1$ and $\dot{u}_2$ as "mechanical currents". Thus, an applied magnetic field produces a"mechanical voltage" via the magneto-mechanical coupling factor φ_m and due to the electro-mechanical coupling factor φ_p,

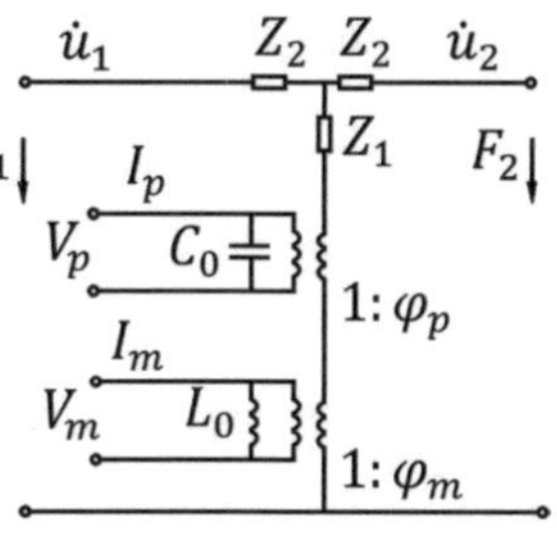

Figure 3.
Magneto-mechano-electric equivalent circuit.

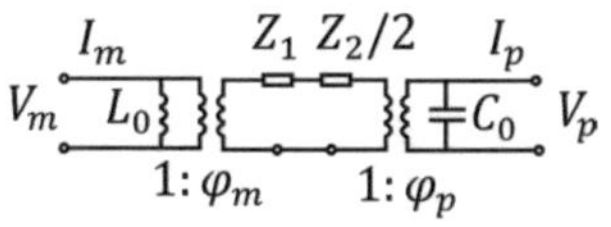

Figure 4.
Magneto-mechano-electric equivalent circuit under condition of free edges.

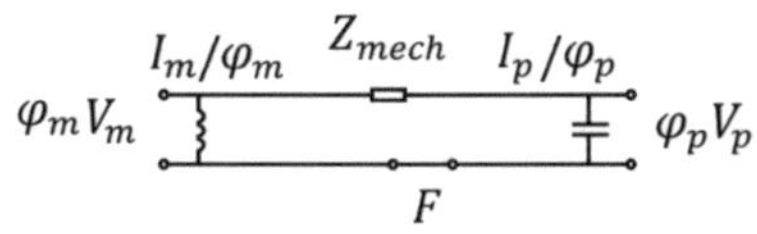

Figure 5.
Magneto-mechano-electric equivalent circuit under free edge condition simplified.

generates an electrical potential difference across the piezoelectric layer. In the circuit model of **Figure 4**, a transformer with ratio $1:\varphi_p$ is used to represent the electro-mechanical coupling. Similarly, an applied electric field due to the magneto-mechanical coupling factor φ_m, generates a magnetization in the magnetostrictive layer. In the circuit model of **Figure 4**, a transformer with ratio $1:\varphi_m$ is used to represent the magneto-mechanical coupling. Note that, the equivalent circuit model considers both electro-mechanical coupling and magneto-mechanical coupling.

Under free-free boundary conditions, that is, null forces at $x = 0$ and $x = \ell$ $(F_1 = F_2 = 0)$, the input and output terminals are grounded, and the circuit is simplified to the one shown in **Figure 5** and then simplified to that shown in **Figure 3**.

The mechanical impedance is: $Z_{mech} = Z_1 + Z_2/2$. Therefore, from Eq. (49) and (50), is obtained:

$$Z_{mech} = -\frac{j}{2}\bar{\rho}\,\bar{v}\,\omega t \cot\frac{k\ell}{2} \tag{53}$$

As the equivalent ME circuit obtained (**Figure 3**) contains symmetrical parts (magnetic and electrical), is possible to discuss both direct and converse effects. To analyze these effects, "open-circuit" and "short-circuit" conditions are: imposed:

Converse ME effect: Under open circuit condition $I_p = I_m = 0$, the following relationship is derived:

$$\frac{\varphi_m dV_m}{\varphi_p dV_p} = \frac{j\omega L_0 {\varphi_m}^2}{j\omega L_0 {\varphi_m}^2 + Z_1 + Z_2/2} \tag{54}$$

Taking into account that: $V_m = -Nj\omega\omega t_m B_3$ in the solenoid and $V_p = t_p E_3$ in the piezoelectric vibrator, and according to Eq. (2), the converse ME coefficient is expressed as:

$$\alpha_{CME(T-T)} = \frac{\partial B}{\partial E} = -\frac{\frac{d_{31,m}}{s_{11}^H \mu_{33}^T}\frac{d_{31,p}}{s_{11}^E} r_p}{\left(\frac{d_{31,m}}{s_{11}^H \mu_{33}^T}\right)^2 r_m + \frac{\omega\ell}{2\mu_{33}^T}\bar{\rho}\,\bar{v}\cot\frac{k\ell}{2}} \tag{55}$$

By indicating α_{CME} with the subscript T–T, it has been possible to omit the subscripts of B and E.

Direct ME effect: Under short-circuit condition $V_p = V_m = 0$, the following relationship is derived:

$$\frac{dI_p/\varphi_p}{dI_m/\varphi_m} = \frac{\frac{1}{\varphi_p{}^2/j\omega C_0 + Z_1 + Z_2/2}}{\frac{1}{j\omega L_0 \varphi_m{}^2} + \frac{1}{\varphi_p{}^2/j\omega C_0 + Z_1 + Z_2/2}} \tag{56}$$

Taking into account Eq. (28) and (37), and according to Eq. (1), the direct ME coefficient is obtained as:

$$\alpha_{DME(T-T)} = \frac{\partial P}{\partial H} = \frac{\frac{d_{31,m}}{s_{11}^H \mu_{33}^T} \frac{d_{31,p}}{s_{11}^E} r_m}{\left(\frac{d_{31,m}}{s_{11}^H \mu_{33}^T}\right)^2 r_m + \frac{r_p}{\varepsilon_{33} \mu_{33}^T}\left(\frac{d_{31,p}}{s_{11}^E}\right)^2 + \frac{\omega \ell}{2\mu_{33}^T} \overline{\rho}\, \overline{v} \cot \frac{k\ell}{2}} \tag{57}$$

By indicating α_{DME} with the subscript T–T, it has been possible to omit the subscripts of P and H.

Thus, expressions of CME and DME coefficients with physical parameters of the phases that constitute the composite material are obtained. They are mathematical expressions that no longer involve circuit parameters but physical parameters of the ferrite and ferroelectric. Eq. (55) and (57) show the expression of the CME and DME coefficients as a function of frequency, however for frequencies much lower than the resonance frequency of the material ($\omega \ll \omega_r$), it can be approximated that $\cot {}^{k\ell}\!/_2 \sim {}^2\!/_{k\ell}$ and as a result the CME and DME coefficients become independent of the frequency and the length of the laminate.

7. ME laminate under resonance frequency

When the alternating field of excitation (electric or magnetic) has a frequency near to the resonance frequency of the material, Eq. (55) and (57) give an infinite value of ME coefficients, which is physically impossible. To predict the ME behavior near to its resonance frequency, it is necessary to consider the effect of mechanical dissipation in the energy conversion process. The mechanical impedance in the resonant state can be approximated as a series RLC circuit shown in **Figure 6**.

For such a model, the circuit impedance is obtained simplify Eq. (53) by using the Taylor series expansion [9]:

$$Z_{mech} = R_{mech} + j\omega L_{mech} + \frac{1}{j\omega C_{mech}} \tag{58}$$

where R_{mech}, L_{mech}, and C_{mech} are the effective mechanical resistance, inductance and capacitance, respectively, which have the expressions:

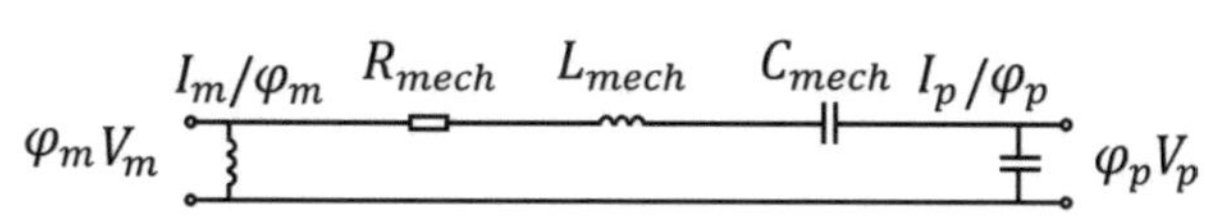

Figure 6.
RLC equivalent circuit under resonance frequency.

$$R_{mech} = \frac{\pi Z_0}{8Q_{mec}} \tag{59}$$

$$L_{mech} = \frac{\pi Z_0}{8\omega_r} \tag{60}$$

$$C_{mech} = \frac{8}{\pi\omega_r Z_0} \tag{61}$$

being $Z_0 = \overline{\rho}\,\overline{\nu}wt$; $\omega_r = {}^{\pi\overline{\nu}}/_{\ell}$ the average resonance frequency of the composite material (note that it holds: $\omega_r{}^2 = L_{mech}{}^{-1}C_{mech}{}^{-1}$) and Q_{mech} the mechanical factor (numerically equal to the inverse of the tangent of mechanical losses). As the mechanical losses in the composite material are the result of the mechanical losses in the magnetic phase ($1/Q_m$) and piezoelectric phase ($1/Q_p$), it can be expressed:

$$\frac{1}{Q_{mech}} = \frac{r_m}{Q_m} + \frac{r_p}{Q_p} \tag{62}$$

Therefore considering mechanical dissipation, the expressions for the CME and DME coefficients are obtained according to:

$$\alpha_{CME(T-T)} = -\frac{\frac{d_{31,m}}{s_{11}^H\mu_{33}^T}\frac{d_{31,p}}{s_{11}^E}r_p}{\left(\frac{d_{31,m}}{s_{11}^H\mu_{33}^T}\right)^2 r_m + \frac{\pi\overline{\rho}\,\overline{\nu}\ell\omega_r}{8\mu_{33}^T}\left(1-\frac{\omega^2}{\omega_r{}^2}\right) + \frac{j\pi\overline{\rho}\,\overline{\nu}\omega\ell}{8\mu_{33}^T Q_{mec}}} \tag{63}$$

$$\alpha_{DME(T-T)} = \frac{\frac{d_{31,m}}{s_{11}^H\mu_{33}^T}\frac{d_{31,p}}{s_{11}^E}r_m}{\left(\frac{d_{31,m}}{s_{11}^H\mu_{33}^T}\right)^2 r_m + \frac{r_p}{\varepsilon_{33}\mu_{33}^T}\left(\frac{d_{31,p}}{s_{11}^E}\right)^2 + \frac{\pi\overline{\rho}\,\overline{\nu}\ell\omega_r}{8\mu_{33}^T}\left(1-\frac{\omega^2}{\omega_r{}^2}\right) + \frac{j\pi\overline{\rho}\,\overline{\nu}\omega\ell}{8\mu_{33}^T Q_{mec}}} \tag{64}$$

From Eq. (63) and (64), it can be seen that the CME and DME coefficients have complex expressions relative to frequency.

8. Simulation of the ME laminate in T: T mode

The behavior of a composite material with lead-free barium titanate (BTO) as ferroelectric component and nickel ferrite (NFO) as ferrimagnetic component is simulated. **Table 1** lists the values used in the model for polycrystalline and polarized phases [10–13]. Arbitrarily, a material geometry of length $\ell = 20$ mm has been selected. With Eq. (63) and (64), the resonance profiles of $|\alpha_{CME}|$ and $|\alpha_{DME}|$ were

NFO	$s_{11}^H\left(\frac{m^2}{N}\right)\cdot 10^{-12}$	$d_{31}\left(\frac{m}{A}\right)\cdot 10^{-12}$	$d_{33}\left(\frac{m}{A}\right)\cdot 10^{-12}$	$\mu_{r\,33}^T$	$\rho_m\left(\frac{g}{cm^3}\right)$	Q_m
	6.5	125	−680	400	5	10
BTO	$s_{11}^E\left(\frac{m^2}{N}\right)\cdot 10^{-12}$	$d_{31}\left(\frac{m}{V}\right)\cdot 10^{-12}$	—	$\varepsilon_{r\,33}^T$	$\rho_p\left(\frac{g}{cm^3}\right)$	Q_p
	7.3	−78	—	1345	7.6	65

Table 1.
Physical parameters of barium titanate (BTO) and nickel ferrite (NFO).

obtained as a function of frequency. The programming codes were developed in the Matlab® software.

Figures 7 and **8** show the results of modules of α_{CME} and α_{DME} at frequencies from 20 kHz to 90 kHz, with r_m as a parameter. From these profiles, it is observed that the $|\alpha_{CME}|$ and $|\alpha_{DME}|$ reach their maximum value when they operate near the resonance frequency. In addition, these results show that the ME coupling is strongly dependent on the thickness ratio between piezomagnetic and piezoelectric components, which is poorly reported with experimental data due to the difficulty in preparing the material.

Figure 9 shows the maxima of $|\alpha_{CME}|$ and $|\alpha_{DME}|$ as a function of the magnetostrictive phase content in the composite material. The cases for $r_m = 0$ (without magnetostrictive layer) and $r_m = 1$ (without piezoelectric layer) have been omitted since under these conditions there is no magneto-elasto-electric coupling and $\alpha_{CME} = \alpha_{DME} = 0$. It is observed that, in absolute values, the DME coefficient

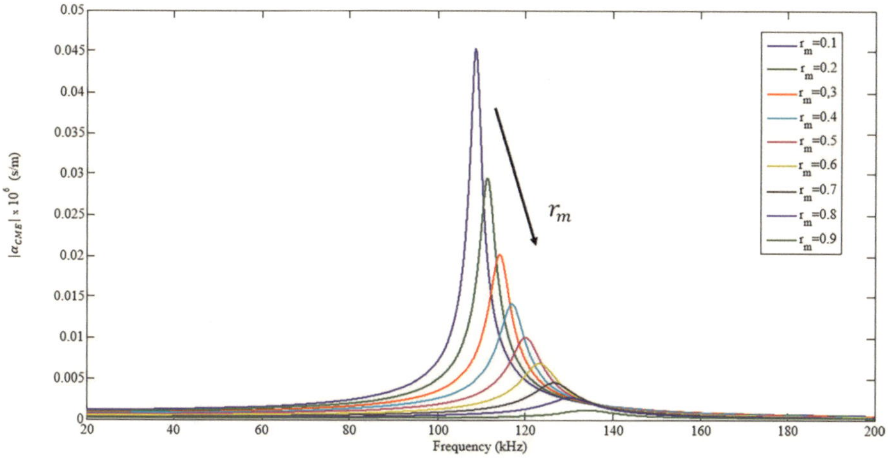

Figure 7.
$|\alpha_{CME}|$ at frequencies from 20 kHz to 90 kHz, for ME laminate in T–T mode.

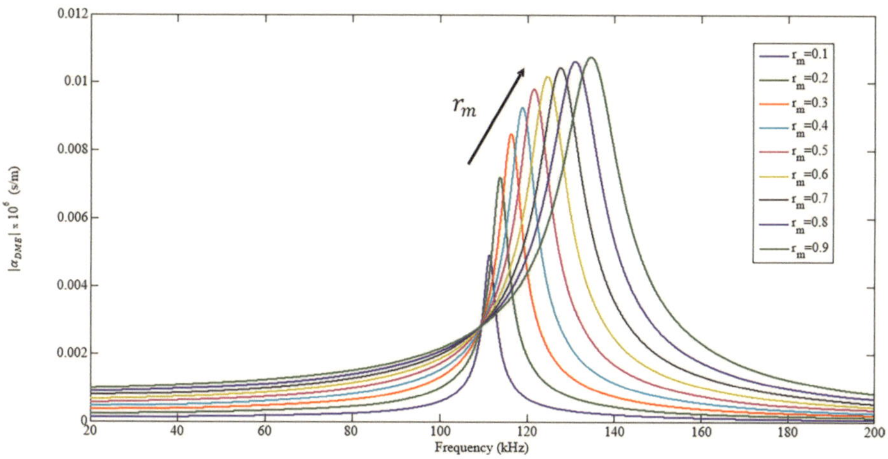

Figure 8.
$|\alpha_{DME}|$ at frequencies from 20 kHz to 90 kHz, for ME laminate in T–T mode.

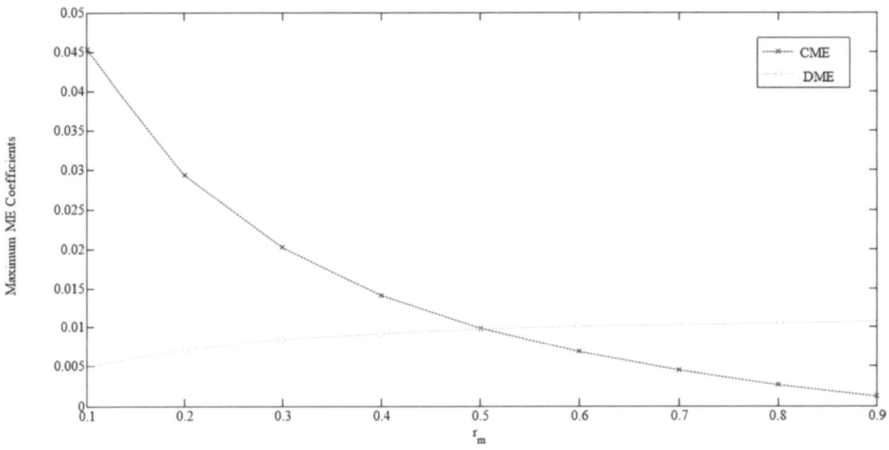

Figure 9.
Maximum of $|\alpha_{CME}|$ and $|\alpha_{DME}|$ versus r_m for ME laminate in T–T mode.

increases with the content of the piezomagnetic component, while the CME coefficient decreases.

In addition, it is observed (**Figure 9**) that $|\alpha_{CME}|$ and $|\alpha_{DME}|$ become similar for composite materials with $r_m \sim 0.5$. This means that for the cases in which the structures have a similar phase content, the energy transition capacity in both effects is similar (the mechanical stress action is mutual).

The resonance frequency of the CME effect can be obtained when the module of α_{CME} reaches its maximum value, so deriving Eq. (63) with respect to the angular frequency ω, the expression is obtained:

$$\omega_{resCME} = \omega_r \sqrt{1 + \left(\frac{d_{31,m}}{s_{11}^{H}\mu_{33}^{T}}\right)^2 \frac{8\mu_{33}^{\overline{T}} r_m}{\pi\bar{\rho}\,\bar{\nu}\ell\omega_r} - \frac{1}{2Q_{mec}^{\ 2}}} \tag{65}$$

Similarly, the resonance frequency of the DME effect can be obtained when the module of α_{DME} reaches its maximum value, so deriving Eq. (64) with respect to the angular frequency ω, the expression is obtained:

$$\omega_{resDME} = \omega_r \sqrt{1 + \left(\frac{d_{31,m}}{s_{11}^{H}\mu_{33}^{T}}\right)^2 \frac{8\mu_{33}^{\overline{T}} r_m}{\pi\bar{\rho}\,\bar{\nu}\ell\omega_r} + \left(\frac{d_{31,p}}{s_{11}^{E}}\right)^2 \frac{8\,r_p}{\varepsilon_{33}\pi\bar{\rho}\,\bar{\nu}\ell\omega_r} - \frac{1}{2Q_{mec}^{\ 2}}} \tag{66}$$

It can be noted that both the CME and DME resonance frequencies are independent of the total thickness of the compound t, they only depend on the thickness ratio between the phases.

Figure 10 shows the resonance frequencies (calculated using the relationship $\omega = 2\pi f$). It is observed that the resonance frequencies (both CME and DME) increase with the ferrite content. In other words, the maxima in the profiles of **Figures** 7 and **8** change at higher frequencies with the magnetostrictive phase content. According to Eq. (65) and (66), the value of ω_{resDME} is always greater than ω_{resCME} due to the term $\left(d_{31,p}/s_{11}^{E}\right)^2\left(8\,r_p/\varepsilon_{33}\pi\bar{\rho}\,\bar{\nu}\ell\omega_r\right)$. This provides a mathematical interpretation for the profiles obtained in **Figure 10**, where it is observed that the DME resonance frequencies are always greater than the CME resonance frequencies, for any value of r_m. As seen in

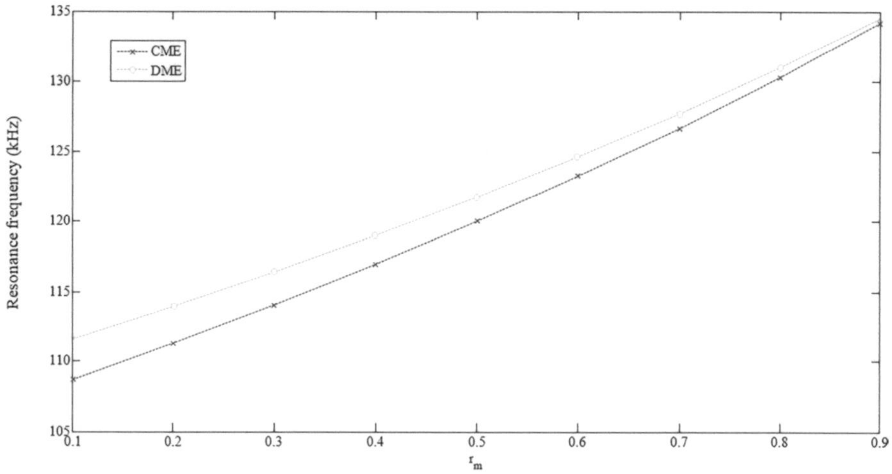

Figure 10.
CME and DME resonance frequencies (f_{resCME} and f_{resDME}) as a function of magnetostrictive phase content, for ME laminate in T–T mode.

Figure 10, the difference between CME and DME resonance frequencies become smaller with the magnetostrictive phase content. That is, $\omega_{resDME} \sim \omega_{resCME}$ for high values of r_m. This allows to infer that the difference in resonance frequencies is mainly caused by the piezoelectric phase.

Another important parameter to characterize the ME effect is the so-called voltage coefficient, defined as:

$$\alpha_V = \frac{\partial V}{\partial H} \tag{67}$$

In the piezoelectric phase is satisfied: $P_3 = \varepsilon_0 \chi_e E_3$, being ε_0 the dielectric permittivity of free space and χ_e the dielectric susceptibility of the material (defined as $\chi_e = \varepsilon_{33}^T/\varepsilon_0 - 1$). Considering that $V_p = t_p E_3$ (Section 5) and that $\left(\varepsilon_{33}^T - \varepsilon_0\right) \sim \varepsilon_{33}^T$ then:

$$\alpha_V = \frac{\partial V_3}{\partial H_3} = \frac{t_p}{\varepsilon_0 \chi_e} \frac{\partial P_3}{\partial H_3} = \frac{r_p t}{\varepsilon_{33}^T} \frac{\partial P_3}{\partial H_3} \tag{68}$$

According to the definition of the DME coefficient (Eq. (1)), the voltage coefficient can be written as:

$$\alpha_V = \frac{r_p t}{\varepsilon_{33}^T} \alpha_{DME} \tag{69}$$

Note that, unlike α_{DME}, the voltage coefficient depends on the thickness t of the material. **Figure 11** shows the results obtained from the voltage coefficients at the resonance frequency (maximum voltage coefficients) as a function of the piezomagnetic phase content (considering a total thickness of $t = 2$ mm in Eq. (69)). To express α_V in units of V/Oe (cgs system), the values in SI units were multiplied by the $10^3/4\pi\left(\frac{V/Oe}{kg\ m^3/C^2\ s}\right)$. It is observed that there is an optimum value of magnetic phase content that corresponds to a maximum value of α_V. More specifically, a maximum of $\alpha_V \sim 80\ mV/Oe$ is observed for $r_m = 0.3$ and at a frequency $f \sim 116\ kHz$.

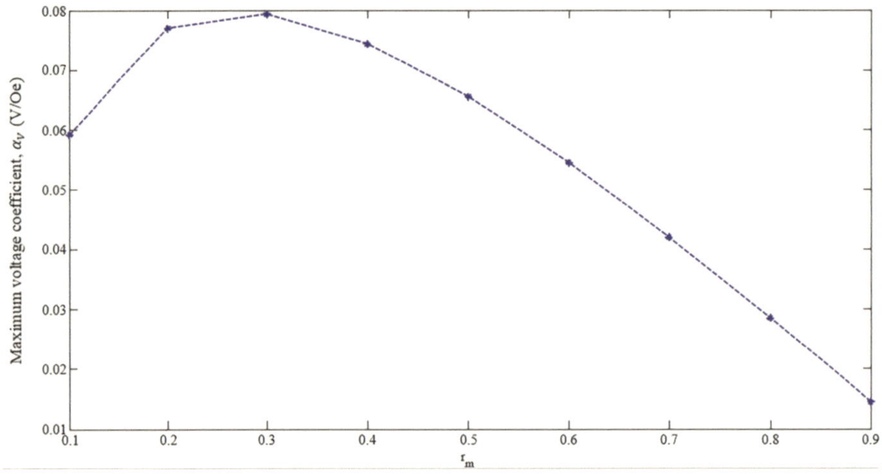

Figure 11.
α_V versus r_m, at resonance frequency, for ME composite in T–T mode.

It is interesting to note that the trend of α_V and α_{DME} with the values of r_m do not coincide with each other. The maximum of α_{DME} is obtained when $r_m = 0.9$ (**Figure 9**), while the maximum in α_V is observed with $r_m = 0.3$ (**Figure 11**). In the field of application, such as transducer devices, most of the time the physical quantity that is measured is the electric potential, so the voltage coefficient could be considered a more significant parameter.

9. Simulation of the ME laminate in L-T mode

If the analysis is extended to an L-T configuration (**Figure 1**), the model must now imply a structure, where the polarization direction of the piezoelectric phase is transverse to the longitudinal direction of the material and the applied magnetic field is longitudinal. **Figure 12a** shows the directions of vectors D (electric displacement), B (magnetic flux density), S (mechanical strain), T (mechanical stress), along with the global coordinates x,y,z. The local coordinates shown in **Figure 12b** are defined in the piezoelectric layer and the local coordinates shown in **Figure 12c** are defined in the piezomagnetic layer.

In L-T mode, the constitutive equations of the piezoelectric material are Eq. (3) and (4) and the constitutive equations of the piezomagnetic material for the strain S_{3m} and the magnetic flux density B_3 are:

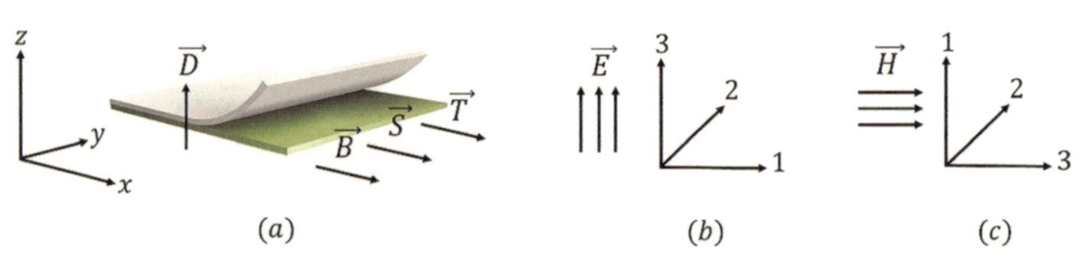

Figure 12.
(a) Schematic illustration of the laminated bilayer ME in L-T mode and local coordinates for the (b) piezoelectric and (c) piezomagnetic layer.

$$S_{3m} = s_{33}^H T_{3m} + d_{33,m} H_3 \tag{70}$$

$$B_3 = d_{33,m} T_{3m} + \mu_{33}^T H_3 \tag{71}$$

Being H_3 the magnetic field intensity in the magnetostrictive layer along the z-direction, T_{3m} the stress in the piezomagnetic layer along the x direction, s_{33}^H the elastic compliance to constant H, $d_{33,m}$ the piezomagnetic constant, and μ_{33}^T the magnetic permeability at constant stress.

Note that in the magnetostrictive layer, the local coordinate "3" (which always designates the direction of magnetization) coincides with the direction of movement, therefore all the subscripts are "3" in the constitutive equations of the magnetostrictive material. In this case, the phases operate as a magnetostrictive transducer in mode 33 and as a piezoelectric transducer in mode 31.

Considering mechanical dissipation, the expressions for the CME and DME coefficients near their resonance frequency became:

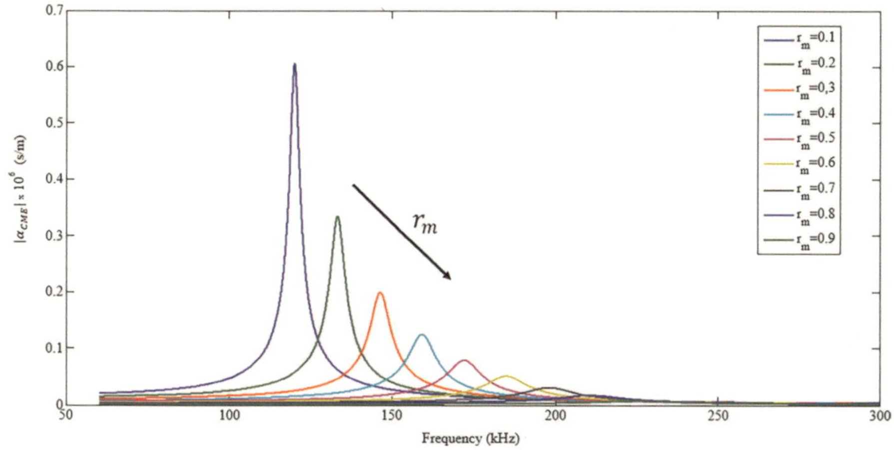

Figure 13.
$|\alpha_{CME}|$ at frequencies from 20 kHz to 90 kHz, for ME laminate in L-T mode.

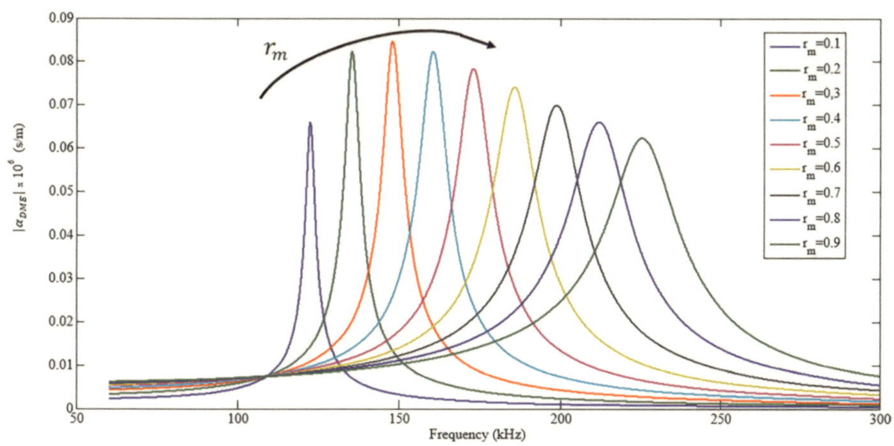

Figure 14.
$|\alpha_{DME}|$ at frequencies from 20 kHz to 90 kHz, for ME laminate in L-T mode.

$$\alpha_{CME(L-T)} = -\frac{\frac{d_{33,m}}{s_{33}^{H}\mu_{33}^{T}}\frac{d_{31,p}}{s_{11}^{E}}r_p}{\left(\frac{d_{33,m}}{s_{33}^{H}\mu_{33}^{T}}\right)^2 r_m + \frac{\pi\bar{\rho}\,\bar{v}\ell\omega_r}{8\mu_{33}^{T}}\left(1-\frac{\omega^2}{\omega_r^2}\right) + \frac{j\pi\bar{\rho}\,\bar{v}\omega\ell}{8\mu_{33}^{T}Q_{mec}}} \tag{72}$$

$$\alpha_{DME(L-T)} = \frac{\frac{d_{33,m}}{s_{33}^{H}\mu_{33}^{T}}\frac{d_{31,p}}{s_{11}^{E}}r_m}{\left(\frac{d_{33,m}}{s_{33}^{H}\mu_{33}^{T}}\right)^2 r_m + \frac{r_p}{\varepsilon_{33}\mu_{33}^{T}}\left(\frac{d_{31,p}}{s_{11}^{E}}\right)^2 + \frac{\pi\bar{\rho}\,\bar{v}\ell\omega_r}{8\mu_{33}^{T}}\left(1-\frac{\omega^2}{\omega_r^2}\right) + \frac{j\pi\bar{\rho}\,\bar{v}\omega\ell}{8\mu_{33}^{T}Q_{mec}}} \tag{73}$$

And the expression for the voltage coefficient in the ME laminate in L-T mode is calculated by means of Eq. (69).

Using Matlab® software to develop the programming codes for the simulation in L-T mode, the results obtained are shown in **Figures 13–17**. The approximation $s_{11}^{H} \sim 3s_{33}^{H}$ was considered in the magnetostrictive layer [8].

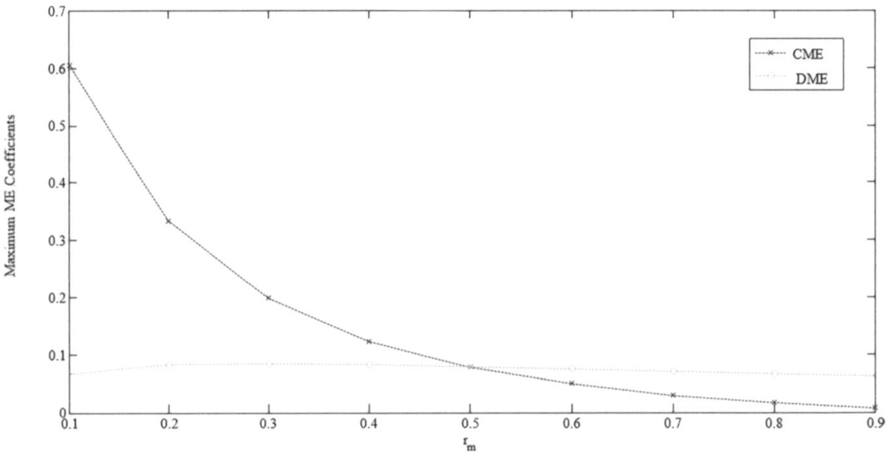

Figure 15.
Maximum of $|\alpha_{CME}|$ and $|\alpha_{DME}|$ versus r_m, for ME laminate in L-T mode.

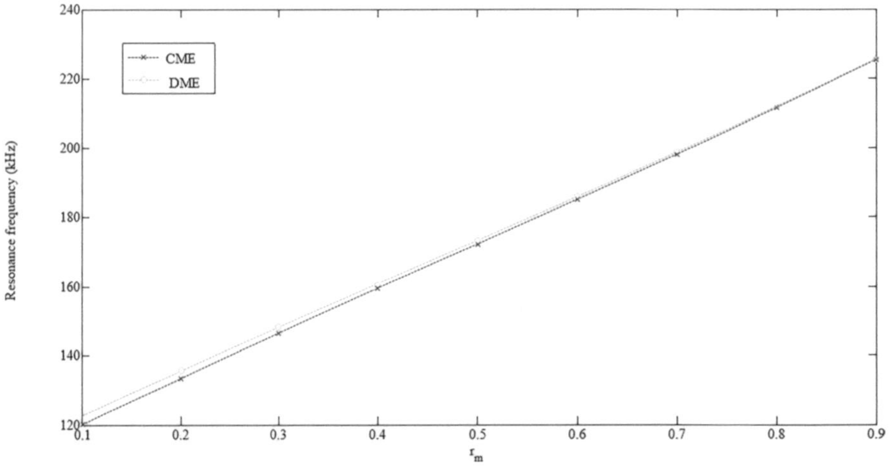

Figure 16.
CME and DME resonance frequencies (f_{resCME} and f_{resDME}) as a function of magnetostrictive phase content, for ME laminate in L-T mode.

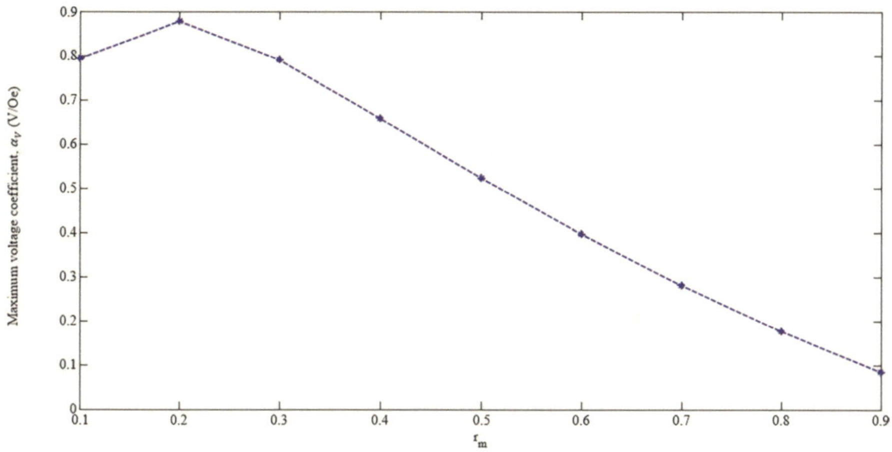

Figure 17.
α_V versus magnetic phase content, in f_{resDME}, for ME laminate in L-T mode.

From the results obtained, it is highlighted that the range of resonance frequencies, both of CME and DME, present a shift towards higher frequencies for the laminate with L-T configuration: the range of resonance frequencies in L-T mode is $\Delta f \sim 120 - 226\ kHz$ (**Figure 16**), while for a T–T configuration it is $\Delta f \sim 108 - 135\ kHz$ (**Figure 10**).

Regarding the ME response analyzed through the voltage coefficient, for the L-T mode a maximum value of $\alpha_V \sim 880mV/Oe$ is found corresponding to an optimal piezomagnetic phase content of $r_m = 0.2$ at a frequency $f \sim 135\ kHz$ (**Figure 17**).

For a better comparison between the results obtained in the L-T and T–T modes, **Figure 18** shows the voltage coefficients α_V (near to resonance frequencies) with the magnetic phase content r_m. It is observed that the responses of the material operating in L-T mode are greater for any magnetic phase content, with respect to the responses in T–T mode. This behavior may be due, at least in part, to energy issues of the ferrite phase. Due to the geometry of the material, the energy component of shape anisotropy presents a lower demagnetization factor in the L-T case, where the magnetization is

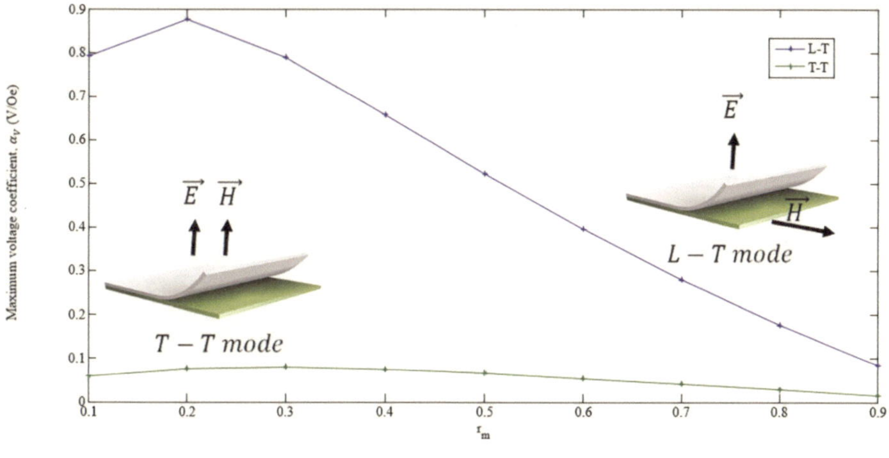

Figure 18.
Maximum α_V versus r_m of the ME laminates, in L-T and T–T modes.

parallel to the length of the material (axis of greatest dimension). This agrees with the trend of the magnetic hysteresis cycles of pure ferrites in bulk, where with measurements in parallel configuration the magnetic responses are greater than those obtained in perpendicular configuration.

The results obtained by applying the circuit model to the barium titanate – nickel ferrite system (Sections 8 and 9) have similar trends to profiles reported by other researchers in analogous systems [14–16]. Additionally, the model was applied in a PZT–Terfenol system [9], observing similar trends to previously published results [17–19]. These similarities in trends allow us to infer that the designed circuit model is useful for modeling the behavior of ME laminated composites. It should be noted that the values of the CME and DME coefficients obtained in the simulation are expected to be somewhat higher than the real values, due to for example the fact that interface effects have not been considered in the model.

10. Conclusions

From the development of magneto-mechano-electric equations, in this chapter it was possible to simulate the direct and converse ME effects of composite laminates by developing an equivalent circuit model, thus predicting the behavior of materials in the presence of a dynamic electromagnetic field.

It is found that the ME material has significantly higher CME and DME coefficients near the resonance frequency. The resonance frequency depends on mode of operation and the phase thicknesses.

For the T–T and L-T modes, the coefficients of the direct and converse effects become similar for materials in which the structures have similar phase content ($r_m \sim 0.5$) suggesting that both effects share the same energy transmission capacity and are relative to the thickness ratio.

We found that the resonant frequency range of CME and DME exhibits a shift towards higher frequencies for the laminate with L-T configuration ($\Delta f \sim 120 - 226\ kHz$) compared to T–T configuration ($\Delta f \sim 108 - 135\ kHz$).

From the results obtained, it is observed that for the barium titanate – nickel ferrite system, the variation of T–T to an L-T mode generates notably higher values of α_V coefficients, in addition to a shift towards higher values of resonance frequency. The maximum voltage coefficient for the T–T mode has a maximum value of $\alpha_V \sim 80\ mV/Oe$ with an optimal piezomagnetic phase content of $r_m = 0.3$ at $f \sim 116\ kHz$. In the L-T mode, the voltage coefficient reaches a maximum value of $\alpha_V \sim 880 mV/Oe$ with an optimal piezomagnetic phase content of $r_m = 0.2$ at $f \sim 135\ kHz$.

The results obtained allow us to infer that, at least in the frequency region studied, the tuning of these frequencies is an important step in the design of these materials for use as transducer devices.

DOI: http://dx.doi.org/10.5772/intechopen.107749

Author details

María Sol Ruiz* and Adrián César Razzitte
Faculty of Engineering, University of Buenos Aires, Ciudad Autónoma de Buenos Aires, Argentina

*Address all correspondence to: mruiz@fi.uba.ar

References

[1] Lottermoser T, Meier D. A short history of multiferroics. Physical Sciences Reviews. 2021;**6**(2):20200032. DOI: 10.1515/psr-2020-0032

[2] Ma J, Hu J, Li Z, Nan CW. Recent Progress in Multiferroic Magnetoelectric composites: From bulk to thin films. Advanced Materials. 2011;**23**(9): 1062-1087. DOI: 10.1002/adma.201003636

[3] Fiebig M. Revival of the magnetoelectric effect. Journal of Physics D: Applied Physics. 2005;**38**(8):R123-R152. DOI: 10.1088/0022-3727/38/8/R01

[4] Ruiz MS, Jacobo SE. Effect of composition on dielectric and magnetic properties of multiferroics and ferroics. IEEE Biennial Congress of Argentina, ARGENCON. 2014:464-467. DOI: 10.1109/ARGENCON.2014.6868536

[5] Zhai JY, Cai N, Shi Z, Lin YH, Nan C-W. Coupled magnetodielectric properties of laminated $PbZr_{0.53}Ti_{0.47}O_3$/$NiFe_2O$ ceramics. Journal of Applied Physics. 2004;**95**(10):5685-5690. DOI: 10.1063/1.1699499

[6] Ikeda T. Fundamentals of Piezoelectricity. Oxford University Press; 1990. p. 263

[7] Nye JF. Physical properties of crystals. In: Their Representation by Tensors and Matrices. Oxford University Press; 1985. p. 352

[8] Engdahl G. Handbook of Giant Magnetostrictive Materials. Elsevier Inc.; 2000. p. 386. DOI: 10.1016/B978-0-12-238640-4.X5014-1

[9] Ruiz MS, Razzitte AC. Equivalent circuit model of magnetostrictive/piezoelectric laminated composite. Global Electromagnetic Compatibility Conference (GEMCCON). 2016:1-6. DOI: 10.1109/GEMCCON.2016.7797315

[10] Ruiz MS, Fano WG, Faigón AN, Razzitte AC. Effect of the inclusion of lead on the dielectric response of doped titanates. Procedia Materials Science. 2015;**8**:641-648. DOI: 10.1016/j.mspro.2015.04.120

[11] Ruiz MS, Bercoff PG, Jacobo SE. Shielding properties of CuNiZn ferrite in radio frequency range. Ceramics International. 2013;**39**(5):4777-4782. DOI: 10.1016/j.ceramint.2012.11.067

[12] Wang T-Z, Zhou Y-H. A theoretical study of nonlinear magnetoelectric effect in magnetostrictive–piezoelectric trilayer. Comp. Struct. 2011;**93**(5): 1485-1492. DOI: 10.1016/j.compstruct.2010.12.003

[13] Chang CM. Coupling Effect of Finite Magneto-Electric Laminate Composites [Thesis]. USA: University of California, UCLA Engineering; 2007

[14] Islam RA, C-b R, Liu JP, Priya S. Effect of gradient composite structure in cofired bilayer composites of Pb $(Zr_{0.56}Ti_{0.44})O_3$–$Ni_{0.6}Zn_{0.2}Cu_{0.2}Fe_2O_4$ system on magnetoelectric coefficient. Journal of Materials Science. 2008; **43**(18):6337-6343. DOI: 10.1007/s10853-008-2912-7

[15] Islam RA, Ni Y, Khachaturyan AG, Priya S. Giant magnetoelectric effect in sintered multilayered composite structures. Journal of Applied Physics. 2008;**104**(4):44103. DOI: 10.1063/1.2966597

[16] Srinivasan G, DeVreugd CP, Hayes R, Bichurin MI, Petrov VM. Magnetoelectric effects in ferromagnetic

and piezoelectric multilayer composites. Magnetoelectric Interaction Phenomena in Crystals. Science Series. 2004;**164**: 35-55. DOI: 10.1007/978-1-4020-2707-9_2

[17] Zhou H-M, Ou X-W, Xiao Y, Qu S-X, Wu H-P. An analytical nonlinear magnetoelectric coupling model of laminated composites under combined pre-stress and magnetic bias loadings. Smart Materials and Structures. 2013; **22**(3):035018. DOI: 10.1088/0964-1726/22/3/035018

[18] Zhou J-P, Wang P, Yang J, Liu P, Zhang H-W. Large converse ME response in Rosen-type transformer and Terfenol-D laminated composite. Journal of Applied Physics. 2012;**111**(3):033915. DOI: 10.1063/1.3684604

[19] Guo M, Dong S. A resonance-bending mode Magnetoelectric coupling equivalent circuit. IEEE Tran. Ultrason. Ferroelectr. Freq. Control. 2009;**56**(11): 2578-2586. DOI: 10.1109/TUFFC.2009.1346

Section 2

Thermoelectric Materials

Chapter 6

Thermal Conductivity in Thermoelectric Materials

Euripides Hatzikraniotis, George S. Polymeris and Theodora Kyratsi

Abstract

Thermal conductivity is a key parameter in identifying and developing alternative materials for many technological and temperature-critical applications, ranging from higher-temperature capability thermal barrier coatings to materials for thermoelectric conversion. The Figure of Merit (ZT) of a thermoelectric material (TE) is a function of the Seebeck coefficient (S), the electrical conductivity (σ), the total thermal conductivity (κ) and the absolute temperature (T). A highly-performing TE material should have high S and σ and low κ. Thermal conductivity has two contributions, the electronic (κ_E) and the lattice (κ_L). Various models have been developed to describe the lattice component of thermal conductivity. In this chapter, the models for the evaluation of lattice thermal conductivity will be explored, both phenomenological as well analytical models, taking into account the various phonon-scattering processes, with examples of real materials.

Keywords: thermoelectric materials, thermal conductivity, analytical and phenomenological models, point defects, disorder, alloying, nano-structuring

1. Introduction

The world's demand for energy is increasing, along with its environmental impact. In fact, the majority of the energy actually produced by man is lost as waste heat. Only an average of about 1/3 of the energy produced by thermal and nuclear power plants, and similarly, only 30% of the energy produced in the internal combustion engine is converted to useful work [1]. If only a portion of the enormous amount of unused waste heat, generated by domestic heating, automotive exhaust and industrial processes could be recovered, it would cause a tremendous benefit in energy demands. Thermoelectricity (TE) could be of great use in energy harvesting [2, 3].

The thermoelectric performance is ranked by the figure of merit ZT ($ZT = S^2\sigma T/\kappa$), where S is the Seebeck coefficient, σ the electrical conductivity, κ the total thermal conductivity and T the absolute temperature. A highly-performing TE material should have high electrical conductivity and Seebeck coefficient and low thermal conductivity.

Thermal conductivity is a key parameter in identifying and developing alternative materials for many technological and temperature-critical applications, ranging from higher-temperature capability thermal barrier coatings to materials for thermoelectric conversion. Since the charge carriers (electrons in n-type or holes in p-type semiconductors) transport both heat and charge, thermal conductivity consists of two additive parts ($\kappa = \kappa_L + \kappa_E$), the lattice ($\kappa_L$) and the electronic ($\kappa_E$) contribution. High ZT requires low total thermal conductivity but high σ simultaneously, so, one of the more popular routes towards improving ZT has been to reduce κ_L [4].

In this book chapter, we shall explore both analytical and phenomenological models for evaluating the thermal conductivity in thermoelectric materials. In this book chapter, we shall start with the basic features and ways for the determination of the lattice thermal conductivity in thermoelectric materials, and we shall proceed with the theoretical foundations of the subject. Next, we will explore various effects that affect the lattice thermal conductivity, namely, the role of defects, of alloying, of disorder and we will conclude with the effect on nano-structuring. Each session starts with the necessary theoretical background and proceeds with application examples of the various aspects of different thermoelectric materials.

2. Thermal conductivity in thermoelectric materials

As previously stated, thermal conductivity has two contributions:

$$\kappa = \kappa_L + \kappa_E = \frac{1}{3} C_V u \lambda + \kappa_E \tag{1}$$

where C_V is the specific heat (at constant volume), u is the speed of sound and λ is the phonon mean free path. All these parameters are in general depend on temperature, and materials characteristics, like the Debye temperature, the mean phonon velocity and parameters which are material dependent, like the mean crystalline size, the density of defects a various scales-lengths, and the 3-phonon scattering parameters. Typically, as can be seen in **Figure 1**, the (lattice) thermal conductivity shows a more or less profound peak at low temperatures and a *1/T* decay due to the 3-phonon scattering. At sufficiently low temperatures (below the Umklapp peak), the lattice thermal conductivity follows the increase of the C_V, with a T^n dependence.

Since grain-boundary scattering dominates at this temperature region, the exponent n for an ideal crystalline solid should be equal to 3 [6]. Any deviation from the

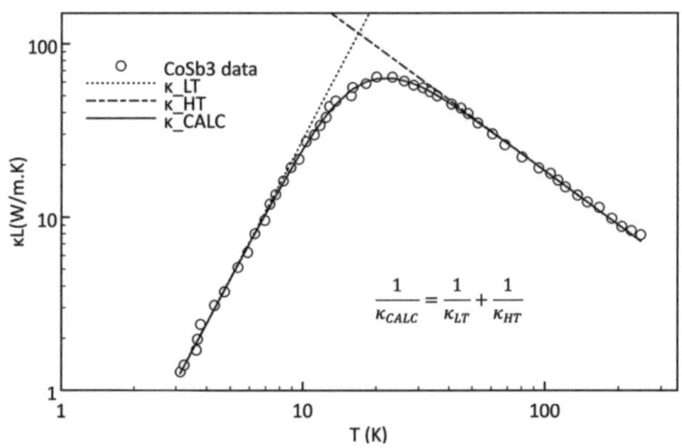

Figure 1.
Lattice thermal conductivity for $CoSb_3$. Data are taken from [5]. The solid line is the phenomenological model. κ_{LT} and κ_{HT} are the descriptions for low temperatures and high temperatures, respectively.

n = 3 at low temperatures, or from the *1/T* dependence at high temperatures may be attributed to additional mechanisms, either to the presence of defects at low temperatures or to radiation losses at high ones.

Assuming that both low- and high-temperature regions, κ_{LT} and κ_{HT}, respectively, can be approximated by a power-law dependence, i.e. $\kappa_{LT} \sim T^n$ and $\kappa_{HT} \sim T^m$, and the lattice thermal conductivity κ_L in **Figure 1**, can be expressed using the phenomenological relation $(\kappa_L)^{-1} = (\kappa_{LT})^{-1} + (\kappa_{HT})^{-1}$ [7], which describes quite well the thermal conductivity κ_L for $CoSb_3$ [5] in the entire temperature range.

2.1 Electronic contribution to thermal conductivity

Figure 2 shows the total thermal conductivity (κ) as a function of temperature for Bi-doped $Mg_2Si_{0.55}Sn_{0.4}Ge_{0.05}$ [8]. As can be seen, as Bi-content increases, doped materials have higher thermal conductivity due to the increased electronic contribution (κ_E). The electronic contribution to thermal conductivity is given using the Wiedemann-Franz law: $\kappa_E = L \cdot \sigma T$, where L is the Lorenz number and σ the electrical conductivity.

Lorentz number L spans from 1.44 to 2.44 $\times$ 10^{-8}WΩK^{-2} for the case of non-degenerate electron gas to the fully degenerated, respectively. Several authors use the highly degenerate value (L = 2.44 $\times$ 10^{-8}WΩK^{-2}), resulting in underestimation of the lattice thermal conductivity, which may reach up to 40% [9]. Therefore, careful evaluation of L is critical in characterising enhancements in ZT due to κ_L reduction. Though modern thermoelectric materials are narrow gap semiconductors with non-parabolic multiple bands, many authors assume single parabolic band (SPB) approximation and scattering of the charge carriers by acoustic phonon (APS), which is the most commonly considered mechanism for thermoelectric materials at high temperatures. Withing the SPB-APS consideration, both L and S are functions only of reduced Fermi level (η), and can be used for the evaluation of L ($S \rightarrow \eta \rightarrow L$):

$$L = \left(\frac{k_B}{e}\right)^2 \frac{2F_0(\eta)F_2(\eta) - 4F_1^2(\eta)}{F_0^2(\eta)}, S = \pm\left(\frac{k_B}{e}\right)\left(\frac{2F_1(\eta)}{F_0(\eta)} - \eta\right) \tag{2}$$

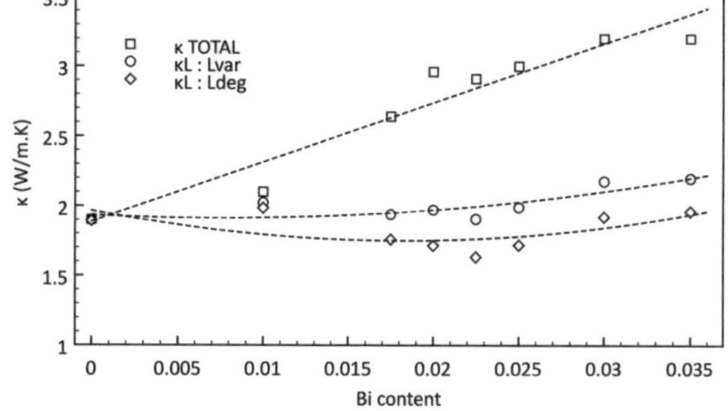

Figure 2.
Total (κ) and lattice thermal (κ_L) conductivity vs. Bi content for Bi-doped $Mg_2Si_{0.55}Sn_{0.4}Ge_{0.05}$. Lines are guide to the eye. Note the underestimation in κ_L if the L = 2.44 $\times$ 10^{-8} WΩK^{-2} (strongly degenerate case) is taken (adapted from [8]).

where $F_j(\eta)$ is the Fermi integral. Kim et al. [10] developed an approximated expression for L as a function of S:

$$L = 1.5 + exp\left(-\frac{|S|}{116}\right) \tag{3}$$

where L is in units of $10^{-8}W\Omega K^{-2}$ and S in μV/K., the approximation of Eq. (3) is far better than considering L = $2.44^{x}10^{-8}W\Omega K^{-2}$. For a comprehensive analysis and the various source or errors for different materials of the evaluation of L, readers may refer to the original paper by Kim et al.

2.2 Bipolar contribution to thermal conductivity

Thermoelectric materials are heavily doped, narrow-gap semiconductors. The high carrier density (typically of the order of 10^{20} cm^{-3}) makes the carrier concentration practically unvaried with temperature. However, since the density of states increases to $T^{3/2}$, the reduced Fermi level gradually drops within the band gap, resulting in the contribution of the minority carriers with increasing temperature. The bipolar (electron-hole pair) thermal conductivity (κ_{BP}) presents a dominant contribution above 500 K in small-bandgap thermoelectrics. According to Glassbrenner and Slack [11], this contribution can be expressed as:

$$\kappa_{BP} = \frac{\beta}{(1+\beta)^2}\left[\frac{Eg}{k_B T} + 4\right]^2\left[\frac{k_B}{e}\right]^2 \sigma T \tag{4}$$

where Eg is the band gap and β is the ratio of electron to hole mobility. Yelgel and Srivastava [12] simplified this expression to:

$$\kappa_{BP} = F_{BP} T^p \, exp\left(\frac{-Eg}{2k_B T}\right) \tag{5}$$

where F_{BP} and p are adjustable parameters, changing with doping type. **Figure 3** shows the subtraction/difference of total and electronic thermal conductivity (κ-κ_E) as a function of the inverse temperature for low Bi-doped $Mg_2Si_{0.55}Sn_{0.4}Ge_{0.05}$ [8]. These

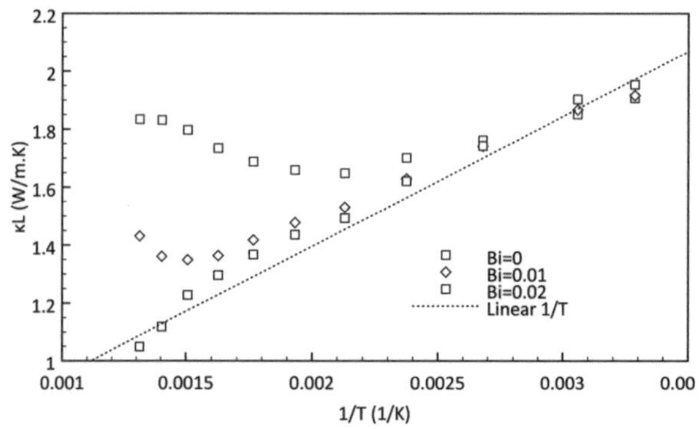

Figure 3.
The effect of bipolar conduction on thermal conductivity. Note the deviation from the linear 1/T tend (adapted from [8]).

values include the bipolar (κ_{BP}) and the lattice contribution (κ_L) of the thermal conductivity. Lattice thermal conductivity exhibits a linear dependence vs. 1/T, and the bipolar contribution is manifested by an upturn at high temperatures (low 1/T) from the linear dependence, as proposed by Kitagawa et al. [13].

At low Bi contents, and at high temperatures, the term κ-κ_{EL} gradually deviates from the linear T^{-1} relationship. This implies that the bipolar diffusion starts to contribute to the thermal conductivity. Suppression of bipolar contribution in thermal conductivity may be achieved either by higher doping or by introducing inclusions of a second phase in the lattice, which act as energy barriers that filters-out the low energy minority carriers. A comprehensive review on how energy barriers act, may be found in [14, 15].

2.3 Theoretical foundations

Advances in thermoelectric technology and materials have led to a considerable amount of computational work aiming at the more accurate simulation of electronic and phononic transport properties for both complex band-structure materials and nano-structured materials. Several techniques have been adopted and have been able to describe electronic transport taking into account the full energy dependencies of the band structure and the scattering mechanisms. They need to merge length scales and physical phenomena, i.e. from atomistic to continuum and from quantum mechanical to fully diffusive, regimes that can co-exist in a typical nanostructured thermoelectric material. They also need to describe transport in arbitrary disordered geometries. A comprehensive book has been recently published in the field of computational modelling and simulation methods for both electronic and phononic transport in thermoelectric materials, and readers should refer to it [16].

Apart from computational simulations, the theory for thermal conductivity is known for several decades now. Though more elaborate models have been introduced, our discussion will be limited to the Debye approximation for specific heat. In order to understand the mechanisms of phonon scattering, semi-classical theoretical calculation based on the modified Callaway's model [17, 18] is considered:

$$\kappa_L(T) = \frac{k_B}{2\pi^2 u}\left(\frac{T}{\Theta_D}\right)^3 \int_0^{\Theta_D/T} \tau_C(x)\frac{x^4 e^x}{(e^x-1)^2}dx$$
$$\kappa_L(T) = \frac{k_B}{2\pi^2 u}\left(\frac{\hbar}{k_B T}\right)^2 \int_0^{\omega_D} \tau_C(\omega)\frac{\omega^4 \exp\left(\frac{\hbar\omega}{k_B T}\right)}{\left(\exp\left(\frac{\hbar\omega}{k_B T}\right)-1\right)^2}d\omega \tag{6}$$

where k_B is the Boltzmann's constant, η is the Plank constant, T is the absolute temperature, u is the average phonon-group velocity, ω_D and Θ_D the Debye frequency and temperature, $x = h\omega/k_B T$ and τ_C is the phonon relaxation time. In the simplest model, the phonon relaxation time combines the scattering from Umklapp processes (τ_U), alloying (τ_A) and grain boundaries (τ_B) according to the Matthiessen's rule [19, 20]:

$$\frac{1}{\tau_C(T,\omega)} = \frac{1}{\tau_B} + \frac{1}{\tau_A(\omega)} + \frac{1}{\tau_U(T,\omega)} \tag{7}$$

some among the aforementioned mechanisms depend only on geometrical features (τ_B), others on phonon frequency (ω) only, while Umklapp on both temperature and phonon frequency. A more elaborated model has been developed by Holland [21] that considers the contribution of thermal conductivity both from the longitudinal and the transverse phonons. It also applies two averaged phonon group velocities to describe the phonon dispersions. Following Callaway and Holland, there have been several modifications to the thermal conductivity model. These modifications are aimed to better capture the temperature dependence over a broader range and focus on achieving a better description of the phonon dispersions rather than developing new thermal conductivity models.

In the simple Callaway model, the various phonon scattering mechanisms may be expressed as follows. Note that the Holland model uses slightly different expressions for the phonon relaxation mechanisms.

Phonon scattering due to interfaces and grain boundaries (τ_B) is given by [18]:

$$\frac{1}{\tau_B} = \frac{4}{3}\frac{u}{L}\frac{1-t}{t} \quad or \quad \frac{1}{\tau_B} = \frac{u}{L_{eff}}, L_{eff} = \frac{3}{4}\frac{t}{1-t}L \tag{8}$$

where t is the phonon transmissivity and L is the average grain size, or scaled to L_{eff}.

When alloying, the solute atom scattering is accounted by the effective medium approach using a Rayleigh-like expression as described by Klemens [22]:

$$\frac{1}{\tau_A(\omega)} = A\,\omega^4, A = \frac{\delta^3}{4\pi u^3}\left[x(1-x)\left\{\left(\frac{\Delta M}{M}\right)^2 + \left(\frac{\Delta\delta}{\delta}\right)^2\right\}\right] \tag{9}$$

where δ is the radius of the solute atom, M is the molar host weight, $\Delta M/M$ is the mass fluctuation, $\Delta\delta/\delta$ is the strain field fluctuation and ε is an adjustable parameter.

The Umklapp scattering, derived from 2nd order perturbation is given by [23]:

$$\frac{1}{\tau_U(\omega,T)} = B_U\omega^2 T exp\left(\frac{-\Theta_D}{T}\right), B_U = \frac{\hbar\gamma^2}{Mu^2\Theta_D} \tag{10}$$

where γ is the Gruneisen parameter and M is the average mass.

Figure 4 depicts the calculated thermal conductivity according to this simple model plotted against the dimensionless T/Θ_D. Two plots are given, increasing the concentration of defects (A and 5A). Thermal conductivity decreases, as expected by increasing the concentration of defects, the peak becomes broader and shifts to slightly higher temperatures, and the temperature dependence of the thermal conductivity in the low-temperature region becomes weaker. The $(T/\Theta_D)^3$ curve is also presented in figure to demonstrate the deviation of the thermal conductivity from the expected ideal T^3 law at low temperatures.

However, the analysis of thermal conductivity is difficult to perform. There are in fact two main complications for accurate analysis; the first is related to the inevitable radiative losses that strongly affect accuracy of measurements at higher temperatures, and the second is related to the numerical problem itself. In order to account for the various mechanisms that limit the lattice contribution to total thermal conductivity, one has to integrate the phonon mean free path over the whole frequency region up to Debye frequency. The numerical problem refers to how to express the phonon mean free path and mean phonon velocity in terms of more readily obtainable or tabulated material parameters. Even in the case of known material parameters, such as the Debye temperature, mean phonon velocity and mean crystalline size, still, the fitting of thermal conductivity data to the integral-based model is not an easy numerical

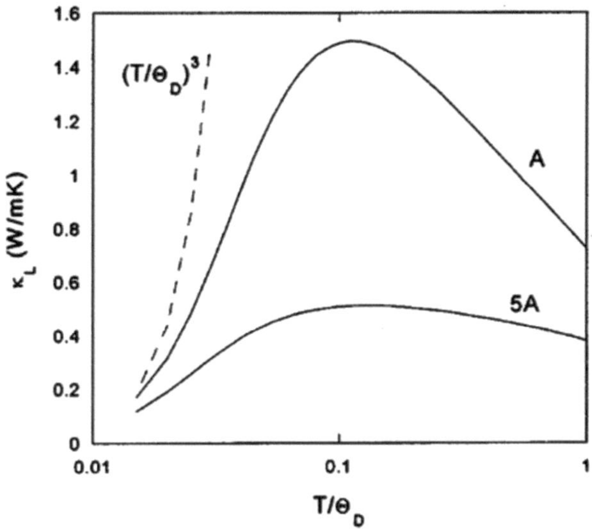

Figure 4.
Theoretical model for lattice thermal conductivity. The two plots correspond to increasing the concentration of defects, A and 5A (after [7]).

problem. Though some authors have performed the fitting to the integral relation 6, or even more complex ones [5, 24], still many authors prefer the adoption of simplified or phenomenological models, which shall be examined in the following.

2.4 The effect of point defects

Defects strongly impact both the electronic and thermal properties of solids, and defect engineering is ubiquitous in the field of thermoelectrics. Therefore, introducing point defects by doping and alloying is the historically most important and robust approach for tuning the charge carrier concentration and reducing κ_L (by scattering high-frequency phonons) [25].

Figure 5 is depicted the inverse of thermal conductivity ($1/\kappa_L$) as a function of the dimensionless T/Θ_D. Data were taken from **Figure 4**, for two cases of increasing defect concentration (A and 5A). As can be seen, the calculated thermal resistivity exhibits a good linear relation with temperature, for temperatures $T > \Theta_D/3$, and the constant term in the linear trend is strongly related to the concentration of defects. The change in the slope is negligible, and this can be attributed to the unvaried value of the Umklapp term B.

In this simple phenomenological model for examining the effect of defects, the observed linear dependence can be understood as follows: The lattice thermal resistivity (w_L) is considered to arise from two contributions according to [26]:

$$w_L = \frac{1}{\kappa_L} = w_o + \Delta w = \alpha T + \beta \tag{11}$$

where the term w_0 is the lattice thermal resistivity of the ideal crystal and Δw represents a contribution due to the presence of defects. For the ideal defect-free crystal $w_0 \sim T$ in the high-temperature range and $\Delta w = 0$. This linear dependence has

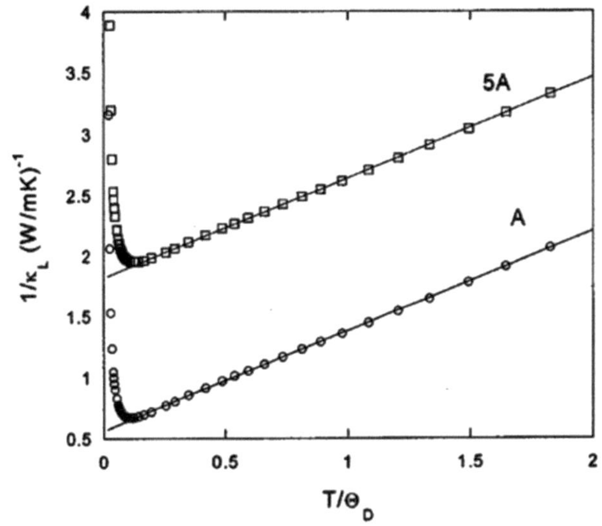

Figure 5.
Calculated lattice thermal resistivity for two different concentrations of point defects A and 5A, respectively (after [7]).

been observed in many thermoelectric materials [7, 26–28] and enables to separate the expected ideal behaviour of a defect-free crystal from that due to the presence of defects.

2.5 The effect of alloying

Solid-solution alloy scattering of phonons is a demonstrated efficient way to reduce lattice thermal conductivity. It may be viewed as a particular case of deliberately introducing defects in the structure, and since the introduced defects alter the local atomic arrangement from the ideal crystal structure, they will intuitively disrupt and affect the charge and thermal transport properties of solids. Similar to the effect of defects, in the case of alloying, the complexity of fitting the fitting of thermal conductivity data to the integral-based model has led to the development of simplified models.

The simplest phenomenological model was developed by Adacci [29], according to which, the composition dependence of thermal conductivity $\kappa\ (z)$ in solid solutions series $A_{1\text{-}Z}B_Z$ can be modelled using the harmonic mean of the thermal conductivities for the two end members κ_A and κ_B, plus an additional 'bowing' term C, which is introduced to account for the mass and strain fluctuation caused by the substitution of the element B into A:

$$\frac{1}{\kappa_L(z)} = \frac{1-z}{\kappa_A} + \frac{z}{\kappa_B} + \frac{z(1-z)}{C} \tag{12}$$

The Adacci model can be easily extended to ternary or quintenary alloys and has been employed by several authors in the evaluation of the bowing factor [7, 30]. **Figure 6** depicts the lattice thermal conductivity for $K_2\ (Bi_{1\text{-}X}Sb_Z)_8Se_{13}$ solid solutions vs. composition at different temperatures. Dashed lines are from Adacci model. As can be seen, the lattice thermal conductivity in the solid solutions is significantly lower

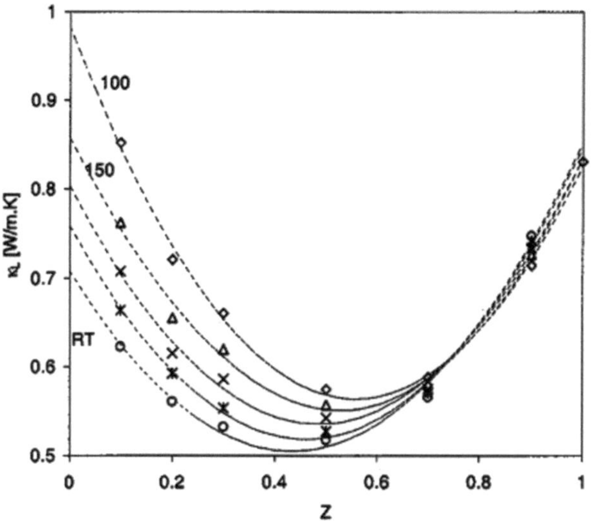

Figure 6.
Lattice thermal conductivity for K_2 $(Bi_{1-X}Sb_X)_8Se_{13}$ solid solutions vs. composition at different temperatures. Dashed lines are from Adacci model (after [7]).

than at the end members, and the incorporation of bowing factor C describes very well the observed compositional dependence.

A more elaborated model has been developed by Klemens [31], which takes into account the mass difference and strain introduced by the alloyed material and the 3-phonon Umklapp process and predicts the ratio of κ_L for a solid solution to that of a hypothetical solid solution without mass or strain fluctuation (κ_O), as a function of a disorder parameter u:

$$\frac{\kappa_L}{\kappa_0} = \frac{\tan^{-1}(u)}{u}, u^2 = \frac{\left(6\pi^5\Omega^2\right)^{1/3}}{2k_B V_S}\kappa_0\Gamma, \Gamma = \frac{\left\langle\overline{\Delta M^2}\right\rangle}{\left\langle\overline{M}\right\rangle^2} + \varepsilon\frac{\left\langle\overline{\Delta R^2}\right\rangle}{\left\langle\overline{R}\right\rangle^2} \tag{13}$$

where Ω is the average volume per atom, V_S is the average speed of sound and the scattering parameter ($\Gamma = \Gamma_M + \Gamma_S$) is related to the average variance in atomic mass (ΔM) and atomic radius (ΔR), with ε an adjustable parameter related to the Gruneisen parameter. For a detailed methodology on the evaluation of Eq. (13) readers may refer to [32]. In general, there will be several different types of atoms that substitute atoms in different sublattices. For the evaluation of the average variance in atomic mass, one should take into account that the kth atom of the ith sublattice has mass $M^k{}_i$, radius $r^k{}_i$ and fractional occupation $f^k{}_i$ [33]. For example, the half-Heusler ZrNiSn has 3 sublattices where substitution may occur (namely the Zr, the Ni and the Sn sublattice), so Sb doping occurs in the Sn sublattice ($ZrNiSn_{0.99}Sb_{0.01}$) while solid solution by Hf may occur in the Zr sublattice (ex. $Hf_{0.5}Zr_{0.5}NiSn_{0.99}Sb_{0.01}$).

The Klemens model uses either only the mass difference (Γ_M) or both the mass and the strain fluctuations (Γ_M and Γ_S). In practice, for a given composition the values of κ_O, Ω and V_S are calculated as linear interpolation between the end-members' values. **Figure 7** depicts the variation of κ_L for $Zr_XTi_{1-X}NiSn$ solid solutions. Results are shown as solid lines, and the smooth variation of Γ_M and Γ_S are shown in the insert. As can be seen, mass fluctuation is a major, but not the only mechanism to describe the

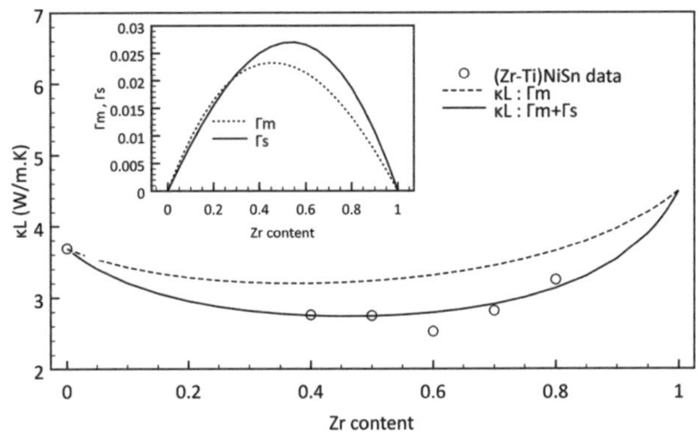

Figure 7.
Evolution of lattice thermal conductivity in $Zr_XTi_{1\text{-}X}NiSn$ (modified from [34]).

composition dependence of the lattice thermal conductivity in $Zr_XTi_{1\text{-}X}NiSn$, and stain is required to model the lattice thermal conductivity at room temperature [34].

2.6 The effect of disorder

The majority of phonon transport has been derived from studies of homogenous crystalline solids, where the atomic composition and structure are periodic. In crystalline solids, the solutions to the equations of motions for the atoms (in the harmonic limit) result in-plane wave modulated velocity fields for the normal modes of vibration. However, it has been known for several decades that whenever a system lacks periodicity, either compositional or structural, the normal modes of vibration can still be determined (in the harmonic limit), but the solutions take on different characteristics and many modes may not be plane wave modulated.

In the case of a highly disordered material, or for non-crystalline solid, Cahill and Pohl [35, 36] developed a model for the temperature dependence of thermal conductivity by adopting, the Einstein model for heat conduction [37] (individual oscillators vibrated independently one from the other) and the Debye model for lattice vibrations, by dividing the material into regions of size $\lambda/2$, which oscillate with frequency $\omega = 2\pi v/\lambda$. This model has been successfully applied to amorphous SiO_2 [36] and many other disordered materials. Similar to Eq. (6), in this case, thermal conductivity κ_D, can be written as [35]:

$$\kappa_D(T) = \frac{k_B}{2\pi^2 u}\left(\frac{\hbar}{k_B T}\right)^2 \int_0^{\omega_L} \tau_D \frac{\omega^4 \exp\left(\frac{\hbar\omega}{k_B T}\right)}{\left(\exp\left(\frac{\hbar\omega}{k_B T}\right) - 1\right)^2} d\omega \qquad (14)$$

where τ_D is the phonon relaxation rate for the highly disordered material, defined as $\tau = \pi/\omega$, and ω_L is the limiting (transition) frequency, much similar defined as ω_D. comparing Eqs. (6) and (14), it can be easily seen that the thermal conductivity in a highly disordered material, κ_D, even in the case where $\omega_L \approx \omega_D$, is substantially lower than κ_L, and results in a monotonous increase, reaching a constant value, $\kappa_{MIN} = \kappa_{Dmax}$, at high temperatures. This is the minimum thermal conductivity.

Eq. (6) does not obviously apply in the case where the phonon mean free path λ becomes of the order of one-half of the phonon wavelength ($\lambda/2 = \pi v/\omega$). Since the phonon means free path, $\lambda = \lambda\ (\omega, T)$, depends on the values of parameters of the various phonon-scattering mechanisms, this situation is plausible even in a crystalline material with high concentration of defects and low phonon-phonon interaction strength. In the general case, the phonon relaxation rate should be [38]:

$$\tau(\omega, T) = \begin{cases} \tau_C(\omega, T), 0 \leq \omega \leq \omega_0 \\ \dfrac{\pi}{\omega} = \dfrac{\lambda}{2}, \omega_0 \leq \omega \leq \omega_D \end{cases} \tag{15}$$

where ω_0 is the transition frequency from crystalline to 'localised' contribution. Since completely localised states do not contribute to thermal conductivity, we assume a 'weak localization' and excitations can hop from one site to the next diffusively, much similar to what Cahill and Pohl had considered in the case of heat conduction in amorphous materials [38].

Figure 8 shows the temperature dependence of thermal conductivity for crystalline $K_2Sb_8Se_{13}$ thermoelectric material. $K_2Sb_8Se_{13}$ has a complex and highly anisotropic structure and morphology, large unit cell, low crystal symmetry and structural site occupancy disorder [39]. As can be seen in **Figure 8**, $K_2Bi_8Se_{13}$ has a very low thermal conductivity with a maximum value ~2.5 W/Km at the low-temperature peak and ~ 0.6 W/Km at the room temperature. It is, therefore, interesting to apply the mixed model presented above in a material with such low thermal conductivity: dashed lines show the contributions of crystalline and localised models, while solid line corresponds to mixed model, which describes the lattice thermal conductivity very well, in the entire temperature range.

In order to evaluate the contributions of the various scattering mechanisms and the transition from the crystalline to the region of 'weak localization', the phonon mean free path, $\lambda(\omega)$, is presented in **Figure 9** as a function of phonon frequency, at 15 K and at 100 K, along with the different scattering contributions. As expected, at low temperatures, at low temperatures $\lambda(\omega)$ is limited by grain size at low frequencies and by defects (impurities) at high, while 3-phonon scattering contributes in the mid-frequency range and is manifested by the Normal process. At 100 K, the contribution

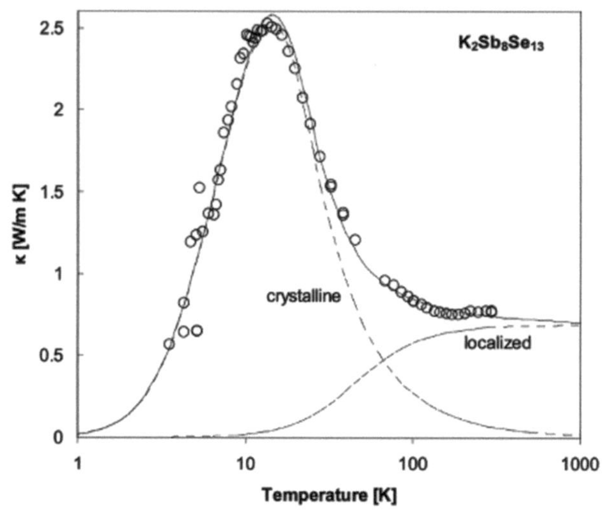

Figure 8.
Temperature dependence of thermal conductivity for $K_2Sb_8Se_{13}$. Dashed lines show the contributions for crystalline and localised models (after 37]).

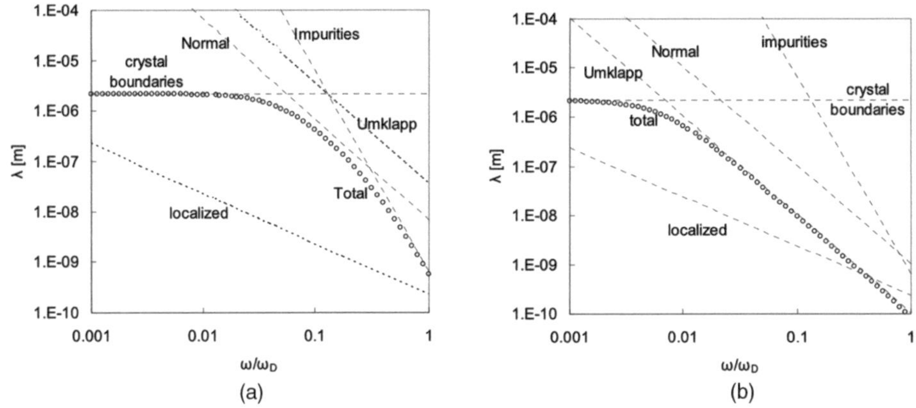

Figure 9.
Calculated phonon mean free path (cirlces) and different scattering contributions as a function of phonon frequency for $K_2Sb_8Se_{13}$ at: (a) 15 K, and (b) 100 K (after [38]).

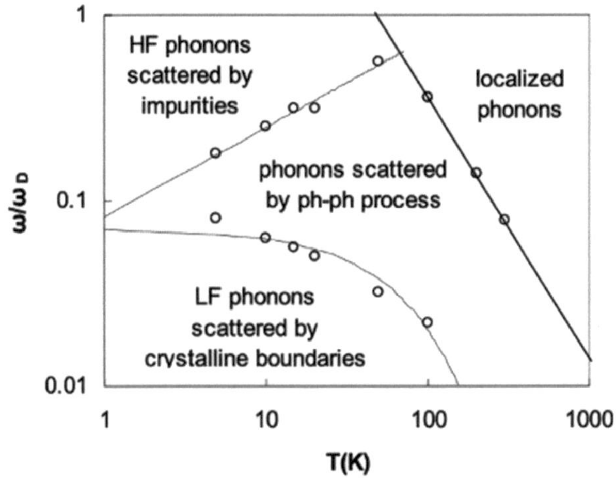

Figure 10.
Summary of phonon processes that limit the lattice thermal conductivity in $K_2Sb_8Se_{13}$ (after [38]).

of grain size is shirked into much lower portion of phonon frequency spectrum, and the Umklapp process prevails over the Normal one. Interestingly, the 'localised' lower dashed curve intersects the phonon mean free path at high frequencies, signalling the transition of the crystalline to the regime of 'weak localization'. In **Figure 10** summarizes the phonon scattering processes that limit lattice thermal conductivity for $K_2Sb_8Se_{13}$ at the entire temperature range: four domains of phonon scattering processes can be assigned within the framework of Debye theory [38].

2.7 The effect of nano-structuring

Nano-structuring has been proved to be an effective way for the reduction of lattice thermal conductivity. Relaxation time of phonon scattering due to the nano-scale inclusions is given by a Mathiessen-type combination of short-wavelength and long-wavelength scattering [40]:

$$\tau_{NP}^{-1}(\omega) = u\left(\sigma_S^{-1} + \sigma_L^{-1}\right)^{-1} N_{NP}, \sigma_S = 2\pi R^2, \sigma_L = \frac{4\pi R^2}{9}\left(\frac{\Delta\rho}{\rho}\right)^2\left(\frac{\omega R}{u}\right)^4 \quad (16)$$

where N_{NP} is the density of the nanoparticles, σ_S and σ_L are the scattering cross-sections for short wavelength (geometrical) and long wavelength (Rayleigh-type), R is the average particle radius, ρ is the medium density and $\Delta\rho$ is the density difference between the particle and matrix materials.

Figure 11 shows the effect of nano-structuring for Mg_2Si (top data) and the combination of alloying/nano-structuring of the pseudo-ternary $(Mg_2Si)_{1\text{-}X\text{-}Y}$ $(Mg_2Sn)_X$ $(Mg_2Ge)_Y$ (bottom). Solid lines represent the theoretical model without (top curve) and with (bottom) alloying, respectively. Dashed lines represent the corresponding models with nano-structuring. Clearly, the experimental points are better described by the lower dashed curve (with nanophases), indicating that an additional phonon scattering mechanism is required.

Figure 12a shows the phonon mean free path (MFP) as a function of the phonon frequency (ω/ω_D), for the various phonon scattering mechanisms encountered. Solid lines refer to RT, dashed at 500 K while the dotted line corresponds to the model without nano inclusions. The various acting mechanisms that are reducing the lattice thermal conductivity are shown in the corresponding figure caption. At low frequencies, phonon MFP is limited by the grain boundary (GB) scattering, resulting in a flat phonon MFP, for frequencies up to $0.01\omega_D$. At high frequencies ($>0.1\omega_D$) phonon MFP is governed by a combination of alloying and Umklapp processes. As only the Umklapp scattering is temperature dependent, this mechanism will squeeze more and more the phonon MFP as the temperature increases, resulting in the decrease of lattice thermal conductivity. In the absence of nano-structuring, the total phonon MFP is smoothly changing from the GB scattering to alloy/Umklapp scattering as shown by the dotted curve. However, when nano-structuring is encountered, a considerable amount of mid-range phonons is scattered by nano inclusions (black curves). The two nano-scattering regimes, the geometrical (σ_S) and the Rayleigh-type (σ_L) are evident. **Figure 12b** shows the effect of concentration of the nano-particles in the phonon MFP at RT. As the concentration of nano-precipitates increases, the geometrical (flat regime) contribution is lowered, and the Rayleigh-type (declining regime) is moving towards lower frequencies. As a result, the nano-inclusions will filter out a more significant portion of the mid-range phonons.

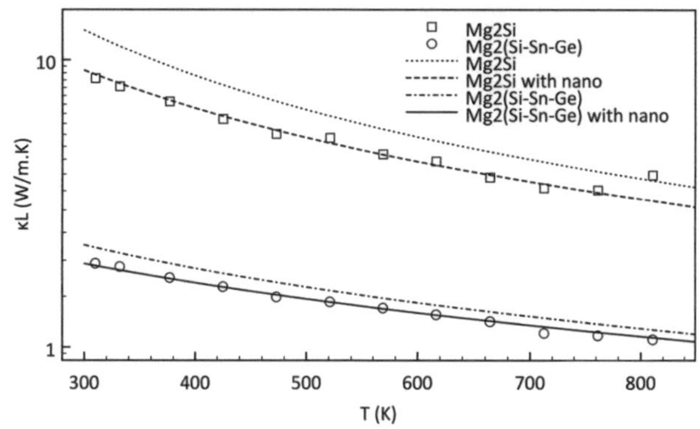

Figure 11.
Lattice thermal conductivity models for Mg_2Si and for $Mg_2Si_{0.55}Sn_{0.4}Ge_{0.05}$. Note the effect of alloying and of nano-structuring in the reduction of lattice thermal conductivity.

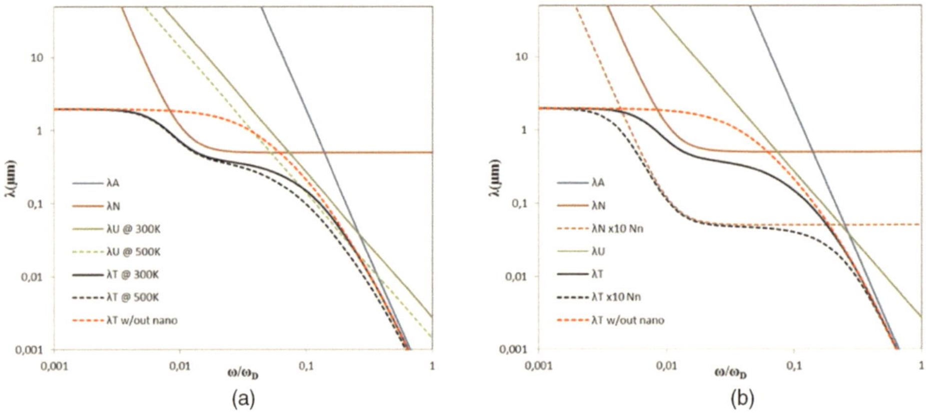

Figure 12.
(a) phonon means free path for $Mg_2Si_{0.55}Sn_{0.4}Ge_{0.05}$ at two temperatures, and (b) the effect of nano-precipitates density.

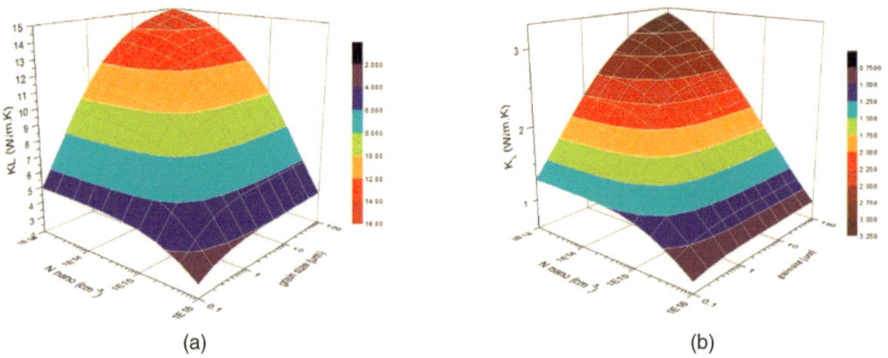

Figure 13.
The interplay of grain size and nano-concentration in the calculated lattice thermal conductivity for Mg_2Si (a) and $Mg_2Si_{0.55}Sn_{0.4}Ge_{0.05}$ (b).

As becomes evident, the effect on nano-structuring has a strong contribution to the geometrical features. Another geometrical-type contribution comes from the grain size (reduced grain size results in reduction of the lattice thermal conductivity). The interplay of the two parameters, namely the grain size and nano-precipitate concentration in the calculated lattice thermal conductivity is depicted in **Figure 13** for Mg_2Si and for $Mg_2Si_{0.6}Sn_{0.4}$. Note that both horizontal axes are in log-scale. Despite the difference in absolute scale, which originated mainly from alloying and the difference in Debye temperature, sound velocity and average mass to a lesser degree, both plots share common characteristics.

Lattice thermal conductivity, as expected, is higher for large grain size and at low nano-precipitate concentration, and lower on the opposite side. The reduction of the calculated κ_L is more than 5 or 4-fold for the cases of the binary and pseudo-binary compounds. Furthermore, the two parameters are interconnected, and for a given grain size there is an optimal value for nano-precipitate concentration, and vice-versa. This optimal pair of parameters lies in the 'diagonal' curve. For values to the left of this optimal-pair curve, lattice thermal conductivity is reduced mainly due to grain size, regardless of value of the nano-precipitate concentration, as indicated by the lines parallel to N-nano axis. For values to the right of the optimal-pair curve, the opposite happens and lattice thermal conductivity is governed by the nano-precipitate concentration (lines parallel to grain-size axis).

This observation may lead to a general approach to the optimization of lattice thermal conductivity in thermoelectric materials, by controlling both the nano- as well as the micro-structural features. For materials grown from melt, nano-structuring may occur using thermodynamically driven phase segregation during cooling of the melt such as spinodal decomposition or nucleation and growth [41]. The size and the concentration of the nano-precipitates can be affected by cooling conditions, however, is difficult to control. Fine-tuning of the grain size, on the other hand, may be effectively done by ball-milling and the control of the milling conditions is much easier [42]. Therefore, tuning the grain size of the resulting compound may give the optimal pair values for the nano- and the micro-structural features.

3. Conclusion

In this book chapter, we have explored both analytical and phenomenological models for evaluating the thermal conductivity in thermoelectric materials. We have started with the basic features and ways for the determination of the lattice thermal conductivity in thermoelectric materials and proceeded with the theoretical foundations of the subject. Next, we have explored various effects that affect the lattice thermal conductivity, namely, the role of defects, alloying, of disorder and we will conclude with the effect on nano-structuring. In each session, we provided both the necessary theoretical background and proceeded with application examples of the various aspects of different thermoelectric materials.

Acknowledgements

E. H. acknowledges the financial support from the project “Design and implementation of innovative lift’s air-conditioning systems by using thermoelectric devices” (Project code: KMP6-0074109) under the framework of the Action “Investment Plans of Innovation” of the Operational Program “Central Macedonia 2014 2020”, that is co-funded by the European Regional Development Fund and Greece. Th.K. acknowledges M-Era.Net project ‘MarTEnergy’ funded by the Cyprus Research and Innovation Foundation (P2P/KOINA/M-ERA.NET/ 0317/04).

Author details

Euripides Hatzikraniotis[1]*, George S. Polymeris[2] and Theodora Kyratsi[3]

1 Department of Physics, Aristotle University of Thessaloniki, Thessaloniki, Greece

2 Institute of Nanoscience and Nanotechnology, National Centre for Scientific Research "Demokritos", Attiki, Greece

3 Department of Mechanical and Manufacturing Engineering, School of Engineering, University of Cyprus, Nicosia, Cyprus

*Address all correspondence to: evris@physics.auth.gr

References

[1] European Environment Agency. Energy and Environment Report No 6/ 2008. Available from: www.eea.europa.eu

[2] Scherrer S, Scherrer H. CRC Handbook of Thermoelectrics. Boca Raton: CRC Press; 1995

[3] Rowe DM. Proceedings on the 9th European Conference on Thermoelectrics. Vol. 4852012

[4] Snyder GJ, Toberer ES. Complex thermoelectric materials. Nature Materials. 2008;7:105

[5] Yang J, Morelli DT, Meisner GP, Chen W, Dyck JS, Uher C. Influence of electron-phonon interaction on the lattice thermal conductivity of $Co_{1-x}Ni_xSb_3$. Physical Review B. 2002;**65**:094115

[6] Bhandari CM, Rowe DM. Thermal Conduction in Semiconductors. New York: Wiley; 1988

[7] Kyratsi T, Hatzikraniotis E, Paraskevopoulos KM, Dyck JS, Shin HK, Uher C, et al. Lattice Thermal Conductivity of $K_2(Bi_{1-z}Sb_z)_8Se_{13}$ Solid Solutions. Journal of Applied Physics. 2004;**95**:4140

[8] Khan AU, Vlachos NV, Hatzikraniotis E, Polymeris GS, Lioutas CB, Stefanaki EC, et al. Thermoelectric Properties of Highly Effective Bi-doped Mg2Si1-x-ySnxGey Materials. Acta Materialia. 2014;77:43

[9] Toberer ES, Baranowski LL, Dames C. Advances in Thermal Conductivity. Annual Review of Materials Research. 2012;**42**:179

[10] Kim H-S, Gibbs ZM, Tang Y, Wang H, Snyder GJ. Characterization of Lorenz number with Seebeck coefficient measurement. APL Materials. 2015;**3**: 041506

[11] Glassbrenner CJ, Slack GA. Thermal Conductivity of Silicon and Germanium from 3°K to the Melting Point. Physical Review. 1964;**134**:A1058

[12] Yelgel ÖC, Srivastava GP. Thermoelectric properties of n-type $Bi_2(Te_{0.85}Se_{0.15})_3$ single crystals doped with CuBr and SbI_3. Physical Review B. **85**:125207

[13] Kitagawa H, Wakatsuki M, Nagaoka H, Noguchi H, Isoda Y, Hasezaki K, et al. Temperature dependence of thermoelectric properties of Ni-doped CoSb3. Journal of Physics and Chemistry of Solids. 2005;**66**:1635

[14] Bahk J-H, Shakouri A. Enhancing the thermoelectric figure of merit through the reduction of bipolar thermal conductivity with heterostructure barriers. Applied Physics Letters. 2014; **105**:052106

[15] Thesberg M, Kosina H, Neophytou N. On the effectiveness of the thermoelectric energy filtering mechanism in low-dimensional superlattices and nano-composites. Journal of Applied Physics. 2016;**120**:23

[16] Neophytou N. Theory and Simulation Methods for Electronic and Phononic Transport in Thermoelectric Materials. Springer; 2020

[17] Callaway J. Model for Lattice Thermal Conductivity at Low Temperatures. Physics Review. 1959;**113**: 1046

[18] Chen G, Zeng T, Borca-Tascius Song TD. Phonon engineering in

nanostructures for solid-state energy conversion. Materials Science and Engineering. 2000;**292**:155

[19] Lo S-H, He J, Biswas K, Kanatzidis MG, Dravid VP. Phonon Scattering and Thermal Conductivity in p-Type Nanostructured PbTe-BaTe Bulk Thermoelectric Materials. Advanced Functional Materials. 2012; **201**:221

[20] He J, Girard SN, Kanatzidis MG, Dravid VP. Microstructure-Lattice Thermal Conductivity Correlation in Nanostructured PbTe0.7S0.3 Thermoelectric Materials. Advanced Functional Materials. 2010;**20**:764

[21] Holland MG. Analysis of lattice thermal conductivity. Physics Review. 1963;**3**(132):2461

[22] Klemens PG. Thermal Resistance due to Point Defects at High Temperatures. Physics Review. 1960;**119**:507

[23] Morelli DT, Heremans JP. Estimation of the isotope effect on the lattice thermal conductivity of group IV and group III-V semiconductors. Physical Review B. 2002;**66**:195304

[24] Chionidis N. Models for the Analysis of Thermal Conductivity. MSc diploma Thesis. Aristotle University of Thessaloniki; 2006 (in Greek)

[25] Hu L, Zhu T, Liu X, Zhao X. Point defect engineering of high-performance bismuth-telluride-based thermoelectric materials. Advanced Functional Materials. 2014;**24**:5211

[26] Lostak P, Drasar C, Krejcova A, Benes L, Dyckc JS, Chenc W, et al. Preparation and some physical properties of tetradymite-type Sb_2Te_3 single crystals doped with CdS. Journal of Crystal Growth. 2001;**222**:565

[27] Smirnov IA, Andreev AA, Kutasov VA. Thermal and Electrical Properties of Sb2Te3: Effect of the Complex Valence Band. Soviet Physics – Solid State. 1970;**11**:2681

[28] Devyatkova ED, Tikhonov VV. Scattering of Phonons and Electrons in Solid Solutions. Soviet Physics – Solid State. 1965;**7**:1427

[29] Adachi S. Lattice thermal resistivity of III–V compound alloys. Journal of Applied Physics. 1983;**54**:1844

[30] Korkosz RJ, Chasapis TC, Lo SH, Doak JW, Kim YJ, Wu CI, et al. High ZT in p-type (PbTe)1-2X (PbSe)X(PbS)X Thermoelectric Materials. Journal of American Chemical Society. 2014; **136**:3225

[31] Klemens PG. The thermal conductivity of dielectric solids at low temperatures (theoretical). Proceedings of the Royal Society of London – Series A. 1951;**208**:108

[32] Gurunathan R, Hanus R, Snyder JG. Alloy Scattering of Phonons. Materials Horizons. 2019:001990

[33] Yang J, Meisner GP, Chen L. Strain field fluctuation effects on lattice thermal conductivity of ZrNiSn-based thermoelectric compounds. Applied Physics Letters. 2004;**85**:1140

[34] Mesaritis G, Ioannou I, Delimitis A, Hatzikraniotis E, Gelbstein Y, Kyratsi T. n-type (Zr, Ti) NiSn half Heusler materials via mechanical alloying: Structure, Sb-doping and thermoelectric properties. Journal of Physics and Chemistry of Solids. 2022;**167**:110735

[35] Cahill DG, Pohl RO. Heat flow and lattice vibrations in glasses. Solid State Communication. 1989;**70**:927

[36] Cahill DG, Watson SK, Pohl RO. Lower limit to the thermal conductivity of disordered crystals. Physical Review D. 1992;**46**:6131

[37] Einstein A. Elementare Betrachtungen über die thermische Molekularbewegung in festen Körpern. Annals of Physics. 1911;**35**:679

[38] Hatzikraniotis E, Chionidis N, Kyratsi T, Paraskevopoulos KM, Kanatzidis MG. On the Analysis of Thermal Conductivity: Models and Applications in K2Sb8Se13. Proceedings of the 3rd European Conference on Thermoelectrics ECT. 2005

[39] Kyratsi T, Chung D-Y, Choi K-S, Dyck JS, Chen W, Uher C, et al. Crystal growth of ternary and quaternary alkali metal bismuth chalcogenides using Bridgman technique. MRS Online Proceedings Library (OPL). 2000;**626**:81

[40] Kim W, Singer SL, Majumdar A, Zide JMO, Klenov D, Gossard AC, et al. Reducing thermal conductivity of crystalline solids at high temperature using embedded nanostructures. Nano Letters. 2008;**8**:2097

[41] Uher C. In: Rowe DM, editor. CRC Handbook of Thermoelectrics: Micro to Nano. Taylor & Francis Group, LLC; 2012

[42] Ioannou M, Hatzikraniotis E, Lioutas C, Hassapis T, Atlantis T, Paraskevopoulos KM, et al. Fabrication of Nanocrystalline Mg2Si via Ball-Milling Process. Powder Technology. 2012;**17**: 523

Chapter 7

Development of Quantum Unit of Temperature Standard in Thermoelectric Research

Svyatoslav Yatsyshyn, Yuriy Bobalo, Tetiana Bubela, Bohdan Stadnyk and Mykola Mykyichuk

Abstract

The quantum standard of temperature based on the revealed quantum unit of the mentioned quantity is studied. It is recommended first to apply as an intrinsic standard. Such a standard does not need permanently recurring measurements against the realization of the SI unit to validate its accuracy. It may be considered as the intrinsic standard of temperature that could be embedded into cyber-physical systems (CPSs) ensuring their precision operation. The methodological base of involvement of the developed standard in the formation of the thermoelectric power of thermoelectric transducers as well as the generator is considered. The feasibility of a unified consideration of the nature of thermoelectric power within macro- and nanothermodynamics is shown. This approach is driven by the increasing use of nano elements based on 1D-, 2D- nanomaterials (nanowires and nanosheets, respectively) and nanostructured materials in technology, in particular, to improve the key parameters of thermoelectric generators and thermometers. In the first case, this is thermoelectric efficiency, and in the second case, the accuracy of thermometers, which is determined by the stability in a time of thermoelectric power.

Keywords: thermoelectricity, quantum standard of temperature, drift of thermo-EMF, elastic strain engineering, thermoelectric transducer

1. Introduction

Thermoelectricity, as it is evident from the papers of International Forum on Thermoelectricity, May 2017, Belfast, is continuously developing in two main areas:

- Thermoelectric phenomena and means for energy production. The main thing here is high parameters: thermoelectric quality factor *ZT* and similar coefficients;

- Thermoelectric phenomena and means for measuring temperature (thermoelectric thermometry). Here the main item seems to be the high stability of thermoelectric characteristics or rather their temperature dependence.

There is a certain gap in the physical mechanisms between these areas, as the goals of both areas are significantly different. Since the second area is more metrological; in our opinion, it is interesting to present its achievements in the perspective of further development of the first area, as more focused on renewable energy and efforts of thermoelectric materials science to improve its energy efficiency. In general, there are good reasons for this: a characteristic feature of modern scientific methodology is interdisciplinary research [1], able to update scientific and technological approaches of the considered thermoelectricity.

2. Promising scientific and technological approaches in thermoelectric materials science

2.1 Elastic strain engineering

Thus, the technology of elastic strain engineering was studied [2] and implemented [3] for multigate field-effect transistor (FET) production. Its cornerstones are the synthesis of nanostructures with inherent elastic effects; applying force and measuring the consequences of its action; study of energy dissipation mechanisms; predicting the results of these effects on specific physical properties (in this case, we are interested in thermo-electromotive force (thermo-EMF)).

Although mechanical failure is a consequence of deformation that should be avoided, elastic deformation can produce a positive effect on the properties of materials. The effect of elastic deformation becomes more obvious at small sizes, because micro/nanoscale materials and structures can withstand exceptionally high deformations until they fail [4]. Studies of elastically loaded nanowires have demonstrated [5] that they can withstand significant elastic deformations, and their bending modulus increases exponentially with decreasing nanowire diameter. At the same time, deformations modify the electronic structure of semiconductor nanowires, causing the metal–insulator transition at room temperature and effectively converting mechanical energy into electrical. Simultaneously, the basin properties substantially depend on the temperature [6] while production and operating of produced means.

2.2 Metrology and thermodynamics

To consider the physical phenomena that alter the main thermoelectric characteristics, we involve the thermodynamics of irreversible processes [7, 8]. The latter had proved its suitability for various fields of science and technology, as it had separated the set of different factors acting on the object (thermoelectric material) into several independent, in the first approximation, groups. First of all, the thermodynamics of massive objects have been developed. According to it, 6 degrees of freedom act on such objects that were the 6 thermodynamic forces and 6 thermodynamic flows. To determine each of them, a system of 6 algebraic equations with 6 unknowns has be solved [9].

In the early stages of the development of thermoelectric thermometry, the determining impact factors were the chemical factors (changes in the content of a components), as well as the corresponding principles of primary transducer's drift due to the composition of thermoelectrodes. Diffusion factors and the corresponding principles of instability (changes in temperature and duration of use) became decisive in the later stages of thermometry, i.e., for protected thermoelectric thermometers.

Mechanical factors and principles of instability have become significant in the further aggravation of the requirements for metrological and operational characteristics of thermometers.

2.2.1 Thermoelectric transducer's conversion functions and chemical impact factor

Thermo-EMF in chemically distributed thermoelectrodes ($grad_x\mu$) in the temperature field ($grad_xT$) is obtained by solving the equation of electric current transfer under the condition that $J_e = 0$:

$$\mathrm{U} = \int_x \alpha[T(x)]\nabla_x T dx + \frac{1}{e}\int_x \nabla_x\mu[T(x)]\,dx \tag{1}$$

Until recently, such thermo electrodes were considered as homogeneous and therefore their impact-function was neglected. That is the well-known equation of thermo-EMF produced in the long and thin thermoelectrode:

$$\mathrm{U} = \int_x \alpha[T(x)]\nabla_x T dx \tag{2}$$

2.2.2 Function of influence at thermo diffusion and functional-gradient thermoelectric materials

The solution of the heat transfer equations from 1273 to 273 K of Fe and Ni impurities with C content and heat transfer Q in molybdenum single crystal leads to the methodological component of the error ΔU_x:

$$\begin{aligned} U = U_0 + \Delta U_X &= \int_T \alpha(T)dT + \frac{QC}{e}\ln\frac{T_H}{T_C} \\ \Delta U_X &= \frac{Q_{Fe}C_{Fe} + Q_{Ni}C_{Ni}}{e}\ln\frac{T_H}{T_C} \end{aligned} \tag{3}$$

Here, the relative chemical function of influence was accessed as the ratio of this function ΔU_x to its value $U_0 : K_X\,(T, C, Q, \dots) = \Delta U_X/U_0 = 0.0026$.

2.2.3 Thermo-EMF and thermodynamics

The basic principles of the thermodynamic method in the phenomenological sense were created in the 1950–1960 on the basis of classical thermodynamics [7] and described in the thermodynamics of irreversible processes [8]. Statistical thermodynamics of nonequilibrium processes is broader as science. Its two parts (thermodynamic and electronic approaches) compose a single doctrine of the properties of material objects, which are manifested in their interactions. In linear thermodynamics, while the system is quite near the equilibrium, the thermodynamic flows J and the force X are conjugated by the relations of Onsager reciprocity:

$$J_i = \sum_j \beta_{ij}\,\mathrm{X}_j\dots\dots(i;j = 1\dots l),\,\partial e\;\beta_{ij} = \beta_{ji}\cdot\dots\dots(i,j = 1\dots l), \tag{4}$$

which are obtained by decomposition of a complex function $J = J(X_1; X_2; \dots X_6)$ in the Taylor series near the point $(X_1; X_2 \dots X_l) \to 0$:

$$J_I(X_1, \dots, X_6) = J_I(0) + \sum\nolimits_J \frac{\partial J_I}{\partial X_J}\bigg|_{(0)} (X_J - 0) + \sum\nolimits_j \sum\nolimits_n \frac{\partial^2 J_I}{\partial X_j \partial X_n}\bigg|_{(0)} (X_J - 0)(X_n - 0) + \dots \tag{5}$$

The basis for considering the transfer processes in the linear approximation provided in conditions $J_I(0) \to 0$ and $X_1; X_2 \dots X_l \to 0$ are:

$$J_I(X_1, \dots, X_6) = J_I(0) + \sum\nolimits_J \frac{\partial J_I}{\partial X_J}\bigg|_{(0)} (X_J - 0) = \sum\nolimits_J \beta_{IJ} X_J \dots \dots (i; j = 1 \dots l) \tag{6}$$

are the results of the study of thermometric substance [10] at the significant, experimentally determined $gradT = 10^4$ K/mm, above which the relationship of thermodynamic forces and flows becomes nonlinear. It is almost impossible to achieve such a significant gradient in industry.

In general, thermoelectric phenomena occur in the presence of thermal and electrical conductivity in a thermodynamic system. That is, in a rather long and thin conductor, which is in a temperature-distributed medium, as a result, thermo-EMF or a value equivalent to that described in connection with it. The corresponding two components of the equation relating to heat flow I_e and charge flow I_T are defined as:

$$I_e = k_1 \left[q^2 E_l - eT\nabla \left(\frac{\mu}{T} \right) \right] - \frac{e}{T} k_2 \nabla T I_T = k_2 \left[qE_l - T\nabla \left(\frac{\mu}{T} \right) \right] - \frac{1}{T} k_3 \nabla T \tag{7}$$

Here E_l is the electric field strength; $k_1; k_2; k_3$ are the kinetic coefficients. From here we can determine the thermo-EMF. As is known, the Seebeck effect consists of the occurrence of the potential difference between open ($I_e = 0$) thermoelectrodes in the cold zone:

$$k_1 \left[q^2 E_l - eT\nabla \left(\frac{\mu}{T} \right) \right] - \frac{e}{T} k_2 \nabla T = 0; E_l = \frac{k_2 - k_1 \mu}{ek_1 T} \nabla T = \alpha \nabla T, \tag{8}$$

here α is a Seebeck coefficient. In this case, thermoelectrodes are considered homogeneous, without a gradient of chemical potential ($\nabla \mu = 0$) along the length at which the temperature gradient ∇T is created. By integrating along the length of the thermocouple located in the temperature gradient zone, we obtain characteristic function or an integrated thermo-EMF of thermocouple:

$$U = \int_x E_l dx = \int_x \alpha_{th} \nabla T dx = \int_T \alpha_{th}(T) dT \tag{9}$$

Here α_{th} is the Seebeck coefficient of thermocouple. The integral, defined by the boundaries $T_1; T_2$ corresponding to the temperatures of reference and hot junctions, allows us to come to the basic laws of thermoelectric circuits.

Previously, we have developed the basics of the thermodynamic approach for estimating the drift of thermo-EMF of the thermocouple. Thermodynamic forces and flows were considered especially in the presence of deformation since due to manufacturing sensitive elements of thermoelectric transducers or the thermoelectric material the mechanical stress and strains can be significant in them. Under the

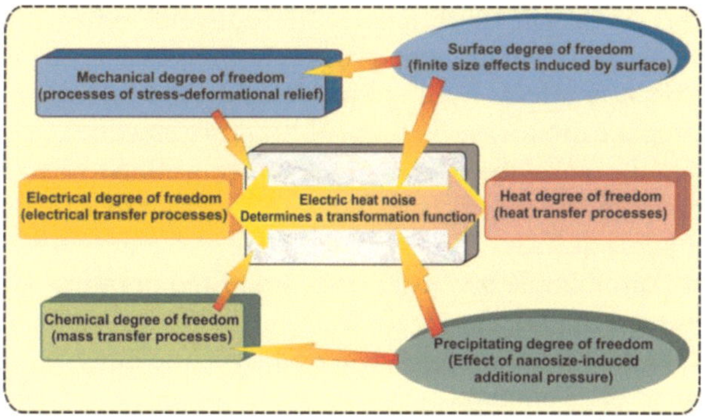

Figure 1.
Conversion function (transfer processes associated with electrical + chemical + heat degrees of freedom that are forming Seebeck coefficient) and influence function (transfer processes associated with the mechanical degree, responsible for drift of thermoelectrodes).

condition of an elastic continuum with dislocations, when $\frac{dU_{int}}{dt}$, the component $p\frac{dV}{dt}$ can be replaced by the product of strain $\hat{\varepsilon}$ and stress $\hat{\sigma}$ (tensors)

$$p\frac{dV}{dt} = \hat{\sigma}\frac{d\,\hat{\varepsilon}}{dt} \tag{10}$$

Phenomenological consideration of stressed thermo electrode shows an increase in Gibbs energy. Then in the basic equation of thermodynamics arises an additional thermodynamic force $X_J = \nabla\left(\frac{\sigma^2}{2E_U}\right) = \frac{\sigma}{E_U \nabla_x \sigma}$ due to this energy, here $\nabla_x \sigma$ is the divergence of mechanical stresses along thin cylindrical thermo electrode with coordinate x. The energy accumulated during deformation significantly affects the conversion function of the transducer by forming a mechanical function of influence (**Figure 1**).

Under the conditions of electric current transfer in the deformed thermoelectric substance of the thermo electrode at sufficiently low temperatures in the absence of diffusion mass transfer and open thermocouple circuit, when $I_e \to 0$:

$$I_e = k_1\left[e^2 E_l - e\frac{\sigma}{mE_U}\nabla\sigma\right] - \frac{e}{T}k_2 \nabla T = 0, \tag{11}$$

It becomes possible to compute the characteristic function U_0 in the form of its nominal value and mechanical impact-function ΔU_M:

$$U(T, \ldots) = U_0(T) + \Delta U_M = \int_x \alpha[T(x)]\nabla_x T dx + \frac{1}{em}\int_x \frac{1}{E_U}\sigma \nabla_x \sigma[T(x)]dx, \tag{12}$$

here E_U is the material's modulus of elasticity.

3. Temperature measurement as a basis for thermoelectric research

3.1 Prerequisites

To characterize the substances of thermoelectric materials science in relation to their properties, regardless of their type, it is necessary to determine exactly the temperature. In thermoelectric thermometry, the conversion function is determined by dependence of thermo-EMF on temperature. In thermoelectric energetics, the temperature significantly affects figure of merit ZT. Independently of the application, the accuracy of temperature measurement [9] becomes essential.

Currently within the world the overall process of transition to a radically higher level of metrology is stimulated by transferring the reference base to a quantum basis. Almost all standards of physical quantities, except temperature, have been replaced [11].

The need for a reproducible quantum standard of temperature was demonstrated by the works presented in [12]. Activities were related to the cardinal problem of thermometry: Committee on Data for Science and Technology has clarified the redefinition of the concept of "Temperature" [13]. It was proposed to replace temperature measurements with energy ones and thus avoid methodological error due to calibration of temperature measuring instruments at the triple point of water. A number of leading metrological centers that are Great Britain (NPL) [14], Germany, etc., have developed a new definition of the unit of temperature: Kelvin, K, is a unit of thermodynamic temperature; its size is determined by fixing the numerical value of Boltzmann constant equal to 1.38065 ... $\bullet\ 10^{-23}$. As a result, the size of the kelvin becomes independent of the material and is determined by the change in thermodynamic temperature, which leads to a change in thermal energy at $k_B T$ equal to 1.380 65 ... $\bullet\ 10^{-23}\ J$. Then it remains to predetermine as accurately as possible the Boltzmann constant, which is realized in [15, 16]. The latter is isolated from other fundamental constants [17]. Therefore, while applying the proposed method as a basic, the obtained results of temperature measurement hold an additional error. It is due to the indirectness of the measurement, which becomes less accurate compared with direct measurement.

This error δT consists of the sum of 2 errors: $\delta E + \delta k_B$. Otherwise, the replacement of temperature measurements inevitably raises a number of difficulties in the field of low-energy measurements related to the sensitivity of devices, insufficient insulation [18, 19] and, most importantly, with establishing the unknown value of the quantum of energy. At the same time, all other standards of physical quantities of the SI system have already become quantum.

3.2 Quantum of temperature and its implementation

The quantum of temperature was first introduced within the system of Planck units in the form of temperature T_P, where it acts as a determining unit of the temperature scale: 0°C = 273.15 K = 1.9279 $\bullet\ 10^{-30}\ T_P$ [16]. We have proved the existence of a quantum of temperature and also revealed the possibility of its use in the quantum temperature standard [20]. The latter is realized on the basis of the current quantum standards of electrical resistance R based on the inversely proportional value of the conductivity quantum ($R = 1/\sigma_Q$) [21] and the voltage standard U based on the array of Josephson junctions [22].

3.2.1 Thermoelectric method of measuring temperature

Independent of temperature, operating current of considered device passes through the material, that is, the superconducting carbon nano tube (CNT) [23], graphene [24], or other substance where the quantum Hall effect may be realized. The design of a CNTFET field-effect transistor with a CNT built into the base is used [25]. When the latter is selected with superconducting properties, such a transistor is characterized by an electrical resistance of 25812.807557 ± 0.0040 Ω [26] caused exclusively by the resistive properties of supply contacts. While conducting current I through the CNT on such contacts, it arises a temperature jump ΔT:

$$\Delta T = \frac{2hI}{3k_{B}e} = \frac{2hN}{3k_{B}\Delta t}, K, \tag{13}$$

when limiting the number of electrons N per unit time to one electron, temperature increase is determined by the ratio of Planck and Boltzmann constants:

$$\Delta T|_{\substack{\Delta t \to 1s \\ N \to 1}} = \frac{2h}{3k_B}\left[\frac{K}{s}\right] \cdot 1[s.] = 3.199493\,42 \bullet 10^{-11}\ \mathrm{K} \tag{14}$$

Thus obtained in such a way, the usniversal value denominated by us [27] as the reduced quantum unit of temperature (RQIT) is independent of the different influence factors and kind of substance. The value of RQUT, being measured in relation to the units of the SI system, is characterized by uncertainty, which is determined by the sum of uncertainties of two constants: Planck constant h and Boltzmann constant k_B [28], which together form its total relative uncertainty equal to $59.2 \bullet 10^{-8}$. For temperature changes of $\sim 10^{-11}$ K due to single-electron relaxation, we must obtain measurable values sufficient to record temperature changes. This means that, first, the current through a carbon nanotube/semiconductor with a quantum Hall effect should be increased.

In this direction, studies conducted [29] found the following. The break of an electric circuit due to the CNT defect can occur when the temperature of the nanotube increases by several hundred degrees, caused, for instance, by the self-heating effect. The latter is less pronounced in semiconductor nanotubes compared with tubes with a metallic conductivity due to different heat dissipation mechanisms. Second, electronic phenomenon with a pronounced integration effect was proposed to register a weak temperature signal. That is the Seebeck effect based on elementary eddy currents [30], which correspond to the cooperative motion of electron groups.

3.2.2 Implementation of the quantum temperature standard

Designing the temperature standard based on the temperature quantum has become possible with unique standards of physical units, including the electrical resistance standard (based on the conductivity quantum inversion) and the electric voltage standard (based on the Josephson junction array) together with the frequency standard. While powering the device from the electric voltage standard, it is possible to pass a certain number of electrons through the electrical resistance standard of nominal value equal to the Klitzing constant $R = h/e^2 = 25812.807 \ldots\ \Omega$). In the considered temperature standard based on the CNTFET [27], the source and drain are made from nickel and copper, which together form a thermocouple through a CNT performing the function of hot quasi-junction of linear dimension ~ 0.02 μm; characteristic function of the mentioned thermocouple is known or predetermined.

As a result, we provide ability to measure the temperature jump on the sensitive element of the standard by thermoelectric method with the minimal methodic error (or with the minimal uncertainty) while controlling the current passing through the CNT. The mode of operation of the device is pulsed. It is powered by a sequence of short ($\sim 10^{-2}$ s) pulses. At the first stage, while supplying, a temperature jump is created. At the second stage (the absence of current), a temperature increase is measured with help of a thermocouple. Specifically, at $I = 10^8$ e/s $\approx 6.24 \bullet 10^{-10}$ A, the temperature jump of $3.2 \bullet 10^{-3}$ K is achieved, which is to be measured. For K-type thermocouple ($\alpha \sim 41$ μV/K), the measured value is 0.14 μV, and for semiconductor thermocouple, it can be up to 1.4 μV. Pre-established uncertainty $\sim 59.2 \bullet 10^{-8}$ permits to suggest that the temperature jump is determined with absolute uncertainty $\sim 1.9 \bullet 10^{-9}$ K.

Subsequent transfer of the set value of the temperature jump from the quantum standard to the reference thermocouple of the first category is carried out by placing its hot junction close to CNT of the temperature standard. If their readouts differ, which may be due to heat loss and other processes, a correction factor should be introduced. The latter may be particularly suitable for studying the thermodynamic processes occurring in the considered standard. That is, this coefficient can be useful for studying not only the processes in thermocouple, but also to clarify the thermodynamic basis of thermoelectricity.

3.2.3 Macro- and nanoproperties related to temperature quantization as the main physical quantity of the SI system

Let us try to link the possibility of temperature quantization with the manifestation of the properties of the bulk level [31]. Naturally, to define it, we use the phenomenon of thermoelectricity, where nanoeffects in the form of a series of elementary eddy currents are involved in the formation of macro characteristics—thermo-EMF [32] due to which we are able to record the small changes in temperature with minimal methodological errors.

4. Development of thermoelectric research and verification of thermoelectricity

4.1 Classical thermodynamics and thermoelectricity

Within Thomson's thermodynamic theory, the Peltier heat is converted into electrical energy with the maximum possible efficiency of the Carnot cycle. Hence, there was obtained the first thermoelectric Thomson ratio ($\pi = \alpha T$), which connects the coefficients of two thermoelectric phenomena: Peltier and Seebeck. The second Thomson thermoelectric ratio ($\tau_A - \tau_B = Td\alpha/dT$), based on the consideration of the heat energy balance in the thermoelectric circuit, has justified the appearance of Thomson coefficients of each of the two thermoelectrodes ($\tau_A; \tau_B$) in the considered thermoelectric circuit as well as firstly predicted thermoelectric effect, contrary to the experimentally proven Peltier and Seebeck effects. In 1853, Thomson verified his own prediction that is the thermodynamic theory of thermoelectricity by recording Thomson's heat with thermometers. In 1867, F. Le Roy had repeated this experiment, replacing thermometers with thermocouples, and confirmed the results of Thomson [33].

With the quantum standard of temperature at our disposal, such studies should be repeated, especially since they relate to nanoscale effects as being performed on nanoobjects (CNT is considered below). Since Thomson consciously realized his discovery of the thermoelectric effect within classical thermodynamics in the absence of irreversibility, and the drift of thermo-EMF is due to irreversible factors (thermodynamics of irreversible processes), then it seems quite interest to conduct research using the quantum standard of temperature.

4.2 Enhancement of the methodology for studying Peltier and Thomson coefficients

To study the Peltier and Thomson coefficients at the nanolevel, with help of the quantum voltage standard, we propose a CNTFET structure in which the source and drain are made of dissimilar materials *A* and *B* composing the hot junction of quasi-thermocouple. Then it becomes possible to calibrate this particular thermocouple and proceed the experiments.

By slightly changing the structure of CNTFET and making the source and drain from the same material, we can study the low-temperature effect of phonon drag on thermo-EMF. After all, it occurs according to [34] at the smallest diameters (1.0 ... 10 nm), which corresponds to the diameter of the CNT of the conductive material in the area of its contact with another bulk.

However, of particular interest are studies of the deviations of the temperature changes of the quantum standard, measured by the additional reference thermocouple, from the predicted values. Otherwise, their dependence on a number of influencing factors helps to identify the peculiarities of the formation of not only thermo-EMF at the micro and nano levels and also to clarify the ambiguous statistical and thermodynamic interpretation of temperature in nanotechnology [35].

Here the approach connected with studying of mechanical stresses impact on thermo-EMF can be interesting. Although mechanical failure is a consequence of deformation that should be avoided, elastic deformation can provide a positive effect on the substance properties. Since this effect becomes more obvious while diminishing, micro/nanoscale materials and structures can withstand exceptionally high elastic deformations until the failure [36].

4.3 Elastic strains engineering and thermoelectric studies

The elastic strains engineering is able significantly improve the defining characteristics of the created standard. The effect of elastic deformation becomes more apparent at small sizes, since micro/nanoscale materials and structures can withstand exceptionally high elastic deformations before failure. Studies of elastically loaded nanotubes have shown the following: (1) Nanotubes can support significant elastic deformations, and their bending modulus increases exponentially as the nanotube diameter decreases; (2) the strains modify the electronic structure of semiconductor nano/micro probes, causing the transition of the metal–insulator at room temperature and effectively converting mechanical energy into electrical energy. The plastic deformation is absent in the nanoscale due to the proximity of the interface surfaces to the location of possible packaging defects (dislocations, nanopores, and nanowires). As a consequence, there is only elastic deformation. The latter have three linear deformation components and three torsional deformation components. Linear deformation is with + and -. Torsional deformations are left and right. In other words, there are 12 types of deformations that can be calculated and reproduced on the one hand and used for dosed and reproducible effects on thermo-EMF while measuring the temperature in the vicinity of CNT attachments. This is the content of the engineering of elastic strains for nanoscale objects of proposed *I-T* converting element, which can be represented as in particular as thermocouple with thermoelectrodes.

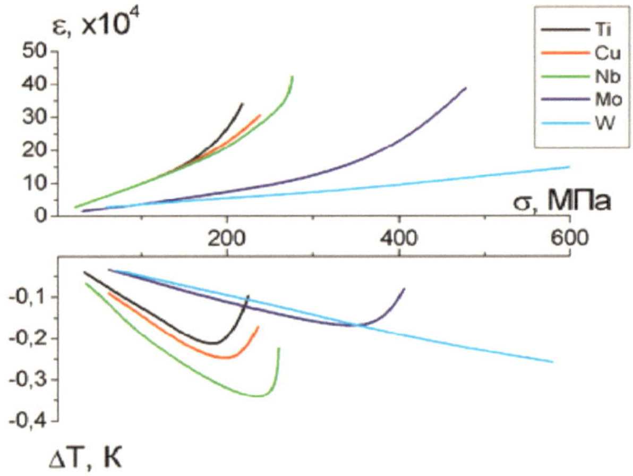

Figure 2.
Influence of elastic deformation on temperature change of pure metals of bulk specimens.

We have studied the influence of elastic deformation (within the limits of up to 1%) on thermo-EMF of different bulk metals and revealed that impact is reversible (**Figure 2**):

Here E_u is Young's module. In particular, for nickel-based alloys, changes in the influence function and the Young's modulus correlate with the coefficient K = 0.598.

The stresses were assessed for the needed current 1 nA passing through superconducting CNT and dissipating in places of CNT attachment, providing the temperature jump ~1.0 K. When the temperature coefficients of linear expansion of the electrode materials (Ni; Cu) are respectively $13{\cdot}10^{-6}\ K^{-1}$ and $16.5{\cdot}10^{-6}\ K^{-1}$, then, at a temperature jump, there arise the thermo structural mechanical stresses. They are caused by the differences of their expansions to the silicon substrate ($5{\cdot}10^{-6}\ K^{-1}$) on which they are fixed. This can lead to an additional uncertainty, which should be taken into account especially for the temperature standard.

4.4 Thermoelectric thermometry and nanothermodynamics

Due to elastic strain engineering, nanostructured materials according to [37] can acquire outstanding opportunities to adjust the physical and chemical properties by changing the field of distributed six-dimensional elastic deformations. For example, in thermoelectric energetics where high efficiency is required, these problems have been mastered designing the generators in 3–4-dimensional space where spiral shapes of thermocouples and thermocouples with directed porosity were invented [38], etc. The same applies to thermoelectric materials.

Thermoelectric thermometry traditionally considers almost all problems in the one-dimensional approximation, which, however, is substantiated by usage of long and thin cylindrical thermoelectrodes. The appearance of functional gradient thermocouples [39] adds one or two more dimensions to solving the drift problems of thermoelectric thermometers. Functionally gradient thermocouples are realized on the basis of thermoelectric materials with smoothly distributed properties (or with a gradient of chemical potential along the electrodes), on which a temperature gradient is imposed during operation. Thus, their conversion function becomes more accurate. Recently, [40] developed an original technology to obtain one-step functionally graded materials fabrication using ultra-large temperature gradients, which may be valuable in the considered issue.

4.5 Nanothermodynamic aspects of updating thermoelectricity

No less important is the study of the correctness of extending macroscopic thermodynamics and statistical physics to nanoscale objects composed of so few atoms that they cannot be described thermodynamically. First of all, this applies to the concept of "temperature" [35]. To transfer the provisions of thermodynamics to nanoscales, it is necessary to understand the unique properties of nanosystems.

4.5.1 Consideration of the action of surface tension force

Maintaining an understanding of metrological approaches to the essence of processes within a thermosensitive substance, we have expanded the range of thermometric methods for studying the role of surface tension gradient as the major thermodynamic force in nanothermodynamics. In particular, we have conducted studies of solid- and liquid-phase sensitive elements while reducing their size in the micro- and nanoarea [41].

In the macroworld, the readouts of a liquid-in-glass thermometer (height of the liquid column Δh), the inner diameter d of which is significant, are described with temperature by the equation: $\Delta h = cd\Delta T$, mm (here c is a constant). As it can be seen, the sensitivity of the thermometer decreases with tube's inner diameter diminishing. The effect of surface tension force is manifested by bending of the meniscus, which leads to the readout error. In the liquid-in-nanotube thermometer is dominated the transfer processes associated with the surface degree of freedom, as the behavior of the liquid in the thin capillary is determined by the effect of wetting dependent on the surface tension force. The temperature dependence of this force is described for different liquids so that the data are embedded in the curve:

$$\sigma V^{2/3} = k(T_c - 6 - T) \qquad (15)$$

Here $k = 2.1 \times 10^{-7}$ J/K mole$^{-2/3}$ is an Eötvös constant; T_C is the characteristic temperature, which in the case of water is equal to 547 K: then the coefficient of surface tension is zero. The Eötvös rule and the Schild equation determine the dependence of the surface tension of the liquid on temperature, and the surface tension coefficient itself is a linear function of temperature. If plot this function, you get the line intersecting the X axis at the characteristic temperature. Equation of calibration characteristics of liquid-in-nanotube thermometer was solved considering the transfer systems for thermosensitive substance with mechanical and surface degrees of freedom, since such thermometer measures the temperature in volume with the insignificant temperature gradient:

$$I_m = -L_{11}\nabla V - L_{12}\nabla M., I_s = -L_{21}\nabla V - L_{22}\nabla M \qquad (16)$$

here $I_m; I_s$ are the liquid transfer flows due to changes in volume and surface tension respectively; L_{ij} are the transfer coefficients. The first Eq. (16) describes the mechanical flow as the flow of displacement of a liquid column under the action of thermodynamic forces due to the gradient of its volume and the gradient of its surface area. The second equation concerns the flow of transfer of the surface degree of freedom that's the flow of alteration in the size of the column under the action of the described thermodynamic forces.

Hence the equation of the calibration characteristic of the liquid-in-nanotube thermometer takes the form (**Figure 3**):

$$\Delta h = b(T_c - 6 - T)/d \tag{17}$$

Here b is a constant.

4.5.2 *Consideration of the action of specific energy in the formation of the precipitations of the second phase in the matrix*

Nanothermodynamics [8] involves the introduction of an additional degree of freedom in the basic equation of thermodynamics. This is γdV (V is the volume of precipitations) due to the loss of specific energy γ while precipitating the second phase in the matrix of the original substance. Issue has been carried out for contact supercooled thin layer of substance at the stage of final sintering of the compressed powder by studying the grains interaction through abovementioned layer. An informative characteristic of the supersaturated phase state was the contact supercooling, which was defined as the difference between the equilibrium temperature of the liquid phase of a given composition with a solid phase at the planar interface and the equilibrium temperature of a crystal with a given surface curvature with the same liquid phase. Supercooling (up to 10 K) of a number of materials was revealed, as a result of which a significant ($\sim$ 10 K/cm) temperature gradient have arisen [42].

4.5.3 *Consideration of additional source of occurrence of eddy currents*

In the classical thermal conductivity, it is believed that the only cause of heat flow in solids is a nonzero temperature gradient. However, the thermal deformation ε of the bulk in the case of intense heat flow leads to the so-called coherence effect, which consists of the interaction of deformation and temperature fields. Evaluation of the effect impact was performed on glass and steel samples, characterizing by the same values of thermal conductivity a_T. The glasses are characterized by the coherence parameter $\frac{\varepsilon'}{a_T T'} < < 400$; for steel, a similar parameter is much lower ($\frac{\varepsilon'}{a_T T'} \sim 20$). This means that for ceramics (nanostructured materials), slight changes in temperature lead to significant consequences of the deformation impact.

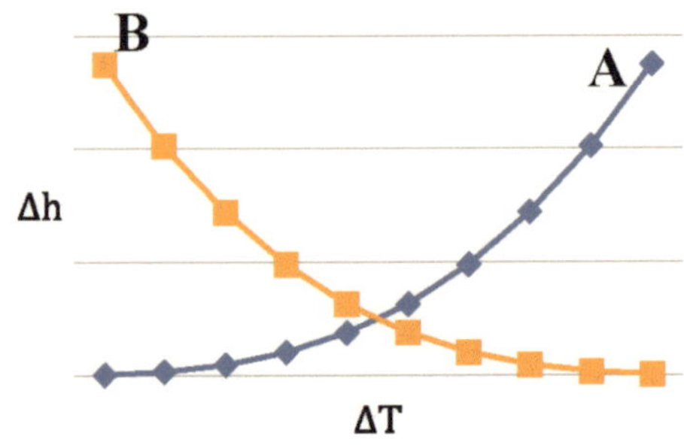

Figure 3.
Calibration characteristic of the liquid-in-glass thermometer (A) and the liquid-in-nanotube thermometer (B).

5. Generation of eddy currents in extreme cases

5.1 Absence of a temperature gradient applied to the thermoelectric material

At one case of thermoelectric material appliance, we can consider the lack of an appreciable temperature gradient imposed on it. Nevertheless, eddy thermoelectric currents can occur and act, because in accordance with the fluctuation-dissipative principles of thermodynamics in the substance, there are fluctuations in thermodynamic parameters, including temperature. Such oscillations are inevitably manifested by eddy thermoelectric currents. The latter lead to the appearance of charge fluctuations on the surface, which can be identified by passive noise spectroscopy [43].

5.2 Significant rate of temperature change in the thermoelectric material

In a substance annealed at the certain temperature and moved to an environment of a higher temperature, due to the formation of dislocation ensembles there are mechanical stresses that can accumulate internal energy. This leads to a change in the noise characteristics, explained by the involvement of the mechanism of energy accumulation-scattering on local quasi-defects of tensile vacancy origin in their interaction with phonons. When current is passed or without it, phonons are generated, accumulated in quasi-defects and then relax in a reversible or irreversible manner, resulting in 1/f or thermal noise [44], respectively. As a result, local eddy thermoelectric currents may occur.

The Raman method makes it possible to study the electron–phonon interaction. The wave number of the optical phonon of the Stokes component significantly depends on the temperature [45]. For example, for monocrystalline silicon, this dependence in the range 300–400 K is linear: $\nu_0(cm^{-1}) = 0.025\Delta T$, here ΔT is the change in temperature of the single crystal. As the temperature increases, the wavelength of the scattered light approaches the wavelength of the laser. This is due to the

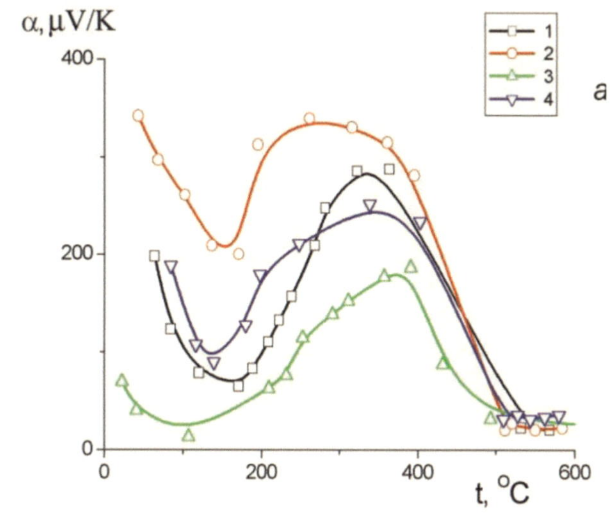

Figure 4.
Seebeck coefficients α of tellurium doped with 1.1% (1) and 1.5% (2) zinc and 1.1% (3) and 1.5% (4) of gallium.

elimination of the tensile action of micro stresses in the tested substance by increasing the mobility of vacancies. Therefore, at pre-melting temperatures, the effect of impurities on thermo-EMF becomes almost the same, as it is evidenced by the behavior of tellurium doped with zinc and gallium while melting above 400°C (**Figure 4**).

6. Conclusion

Progress in thermoelectric energetics is determined by the advances in materials science and is based, on the one hand, on the achievements of nanotechnology as a source of implementation of significant technological trends and, on the other hand, on the achievements of Metrology 4.0 as part of Industry 4.0 [46], which provides the foundation pillars for emerging quantum technologies in the branch of temperature measurements as the main controlled parameter of production and usage of thermoelectric facilities, devices, transducers, and sensitive elements.

It can be accelerated by the introduction of the quantum temperature standard (here the quantum temperature unit is expressed by the ratio of the Boltzmann and the Planck constants), not only as metrological instrument for ensuring the exactness and precision of temperature measurements, and as a tool for studying the spectrum of thermoelectric phenomena by using CNTFET as the kernel construction element, corresponding in the measuring cycle to a nanosized thermocouple. Involvement of the developed elastic strain engineering technology for the manufacture of multigate field-effect transistors gives possibility to improve the efficiency of thermoelectric materials.

Evaluating the achievements of W. Thomson (Lord Kelvin) in thermoelectricity, built on classical thermodynamics, while providing the quantum temperature standard operated on elements based on the thermodynamics of irreversible processes, we gain the ability to study in more detail the reserves of thermoelectricity and to clarify the validity of the basic laws of thermoelectricity for the nanoworld.

Since nanothermodynamics involves the consideration of additional for classical thermodynamics degrees of freedom due to the action of surface tension forces and mechanical stresses, it is possible to modify the thermoelectric properties of materials by forming multidimensional distributed fields of elastic stresses, which can be implemented nowadays. At the same time, nanothermodynamics should develop both in the direction of thermodynamic formalism (introduction of additional surface degree of freedom), which describes the increasingly important role of surface tension forces while reducing the qualificative size of systems to nanodimensions, and in the direction of consideration of elastic stresses in nanoobjects, able to lead to the appearance of anticipations of the second phase. Action of these factors is manifested through thermo-EMF, as a characteristic integrated over the volume of thermoelectric material.

The emergence, study, and subsequent usage of specific mechanisms of eddy thermoelectric currents in thermoelectricity cannot be ruled out. This may be the effect of coherence inherent in nanostructured materials, in which the minimal changes in temperature lead to the associated deformation aftereffect. That is, the influence of deformation on the major thermoelectric parameters has to be meticulously studied. In the future, it is necessary to develop the fundamental principles of generating the eddy thermoelectric currents in the absence/presence of significant temperature gradients imposed on the material. These currents exist in the thermoelectric substance in any case according to the fluctuation-dissipative principles of thermodynamics.

To update the thermoelectricity, the next aspect of the involvement of nanothermodynamics can be considered. That is a justification for the qualified elimination of correlation effects, for which is recommended based on the thermodynamic approach: (a) to select and evaluate independent factors influencing the studied objects; (b) to optimize the number of measurements while providing the necessary metrological characteristics; c) to process the obtained data on the basis of a proper understanding of nanostructured effects as well as substantiating that the detected "artifacts" are not created in the measuring system itself. While studying the measuring instrument as the major unit of the whole system, we have to be sure in its reliability especially concerning the developed thermocouples. It is expedient here to consider thermocouples with built-in self-testing [47]. Moreover, it is recommended by [48] to perform such temperature measurement system basing on thermocouple with controlled temperature field.

Author details

Svyatoslav Yatsyshyn*, Yuriy Bobalo, Tetiana Bubela, Bohdan Stadnyk and Mykola Mykyichuk
Lviv Polytechnic National University, Ukraine

*Address all correspondence to: sviatoslav.p.yatsyshyn@lpnu.ua

References

[1] Ohta M, Jood P, Murata M, et al. An integrated approach to thermoelectrics: Combining phonon dynamics, nanoengineering, novel materials development, module fabrication, and metrology. Advanced Energy Materials. 2018;**9**(23):1801304. DOI: 10.1002/aenm.201801304

[2] Xu N. Elastic Strain Engineering Effectiveness of Strain Solutions for Next-Generation MOSFETs. Berkeley: University of California; 2012

[3] Li J, Shan Z, Ma E. Elastic strain engineering for unprecedented materials properties. MRS Bulletin. 2014;**39**(2): 108-114

[4] Taloni A, Vodret M, Costantini G, Zapperi S. Size effects on the fracture of microscale and nanoscale materials. Na ture Reviews Materials. 2018;**3**:7. DOI: 10.1038/s41578-018-0029-4

[5] Boukai AI, Bunimovich Y, Tahir-Kheli J. Nature. 2008;**451**(7175):168-71. DOI: 10.1038/nature06458

[6] Bernhard IF. Technische Temperaturmessung. In: Physikalische und meßtechnische grundlagen, sensoren und Meßverfahren. Meßfehler: Springer, 2012

[7] Glansdorf P, Prigogine I. Thermodynamic Theory of Structure, Stability and Fluctuations. New York: Wiley; 1971

[8] Geckeler KE, Nishide H, editors. Advanced nanomaterials. John Wiley & Sons; 2009

[9] Yatsyshyn S, Stadnyk B. Accuracy and metrological reliability enhancing of thermoelectric transducers. Sensors and Transducers. 2010;**123**(12):69-75

[10] Ternan J. Thermal drift of thermocouples due to inhomogeneous changes in composition. Journal of Applied Physics. 1984;**55**(1):199-209

[11] de Podesta M. The definition of the Kelvin in the New SI: Its rationale, implementation and implications. In: Proceeding of XIII International Symposium on Temperature and Thermal Measurements in Industry and Science (TEMPMECO 2016); Zakopane, Polska, 26.06-01.07. 2016. p. 12

[12] Yatsyshyn S, Stadnyk B, Ya L, Buniak L. Handbook of Thermometry and Nanothermometry. Barcelona: IFSA Publishing; 2015

[13] Fischer J. International Temperature Scale and New Definition of Kelvin, Meeting on Fundamental Physical Constants, Eltville, 4; 2015

[14] Mills Ia, Quinn T, Mohr P, Taylor B, Williams E. The new SI: Units and fundamental constants, Royal Society Discussing Meeting. 2011

[15] Benz S, Pollarolo A, Qu J, Rogalla H, Urano C, Tew W, et al. An electronic measurement of the Boltzmann constant. Metrologia. 2011;**48**(142):23

[16] Pitre L, Risegari L, Sparasci F, Plimmer MD, Giuliano A. Determination of the Boltzmann constant from the speed of sound in helium gas at the triple point of water. In: Metrologia., 2015;**52**(5):S263. DOI:10.1088/0026-1394/52/5/S263

[17] Lipinsky L, Szmyrka-Grzebyk A. Proposals for new definition of the kelvin. Metrology and Measurement System. 2008;**15**(2):227-234

[18] Hohmann M, Breitkreutz P, Schalles M, Fröhlich T. Calibration of heat flux sensors with small heat fluxes. In: Proceedings of the 58 Internationales Wissenschaftliches Kolloquium: "In Shaping the Future by Engineering". Ilmenau, Germany: Technische Universität; 08-12 Sept. 2014. p. 29

[19] Lindeman M. Microcalorimetry and transition-edge sensor. [Thesis UCRL-LR-142199]; US Department of Energy, Laurence Liverpool National Laboratory; 2000

[20] Yatsyshyn S, et al. Yatsyshyn S, Stadnyk B, editors. Pat. UA № 115601. The method of receiving the temperature quantum on the basis of fundamental constants of the matter and creation of the temperature standard for its realization; 2017 (In Ukr.)

[21] Giesbers A, Rietveld G, Houtzager E, et al. Quantum resistance metrology in graphene. Applied Physics Letters. 2008; **93**:222109-1-3. DOI: 10.1063/1.3043426

[22] A Practical Josephson Voltage Standard at One Volt. Available from: http://www.lee.eng.uerj.br/downloads/graduacao/medidas_eletricasJosephsonJunction.pdf

[23] Schneider B, Etaki S, van der Zant H, Steele G. Coupling carbon nanotube mechanics to a superconducting circuit. Scientific Reports. 2012;**2**:599

[24] Novoselov K et al. Room-temperature quantum hall effect in graphene. Science. 2007;**315**(5817):1379

[25] Sahoo R, Mishra R. Simulations of carbon nanotube field effect transistors. International Journal of Electronic Engineering Research. 2009;**1**(2):117-125

[26] The NIST Reference on Constants, Units, and Uncertainty, CODATA Internationally Recommended 2014 Values on Fundamental Physical Constants. Available from: http://physics.nist.gov/cuu/Constants/index.html

[27] Stadnyk B, Yatsyshyn S. Metrological array of cyber-physical systems. Part 15. Approach to the creation of temperature standard on the basis of fundamental physical constants. Sensors and Transducers. 2016;**199**(4):62-66

[28] Yatsyshyn S, at al. Yatsyshyn S, Stadnyk B, editors. Cyber-physical systems. Metrological Issues. Barcelona: IFSA, International Frequency Sensor Association; 2016. p. 336

[29] Ouyang Y, Guo J. Heat dissipation in carbon nanotube transistors. Applied Physics Letters. 2006;**89**(18):183122. DOI: 10.1063/1.2382734

[30] Luste O, Kuz' R. Computer control of Eddy thermoelectric current. Thermoelectricity. 2004;**2**:11-19

[31] Stadnyk B, Yatsyshyn S. Quantum standard of temperature and possible thermoelectric studies on its basis. Journal of Thermoelectricity. 2017;**2**: 84-96

[32] AL. Chapt 13. William Thomson (Lord Kelvin) and thermoelectricity. In: Collins M, Dougal R, Koenig C, Ruddock I, editors. Kelvin, Thermodynamics and the Natural World. WIT Press; 2015. pp. 337-362

[33] Anatychuk L. Thermoelectricity. In: Thermoelectric Convertors of Energy. Vol. 2. Kyiv-Chernivtsi: Institute of Thermoelectricity; 2003

[34] Baranskii P, Gaidar G. Theory of anisotropy of electron phonon drag thermo EMF in n-Ge. Journal of Thermoelectricity. 2012;**2**:28-37

[35] Yatsyshyn S, Stadnyk B, Ya L. Research in nanothermometry. Part 1. Temperature of micro- and nanosized objects. Sensors and Transducers. 2012; **140**(5):1-7

[36] Penga X, Zhoua L, Lib X, Taob X, Xua G. Strain study of gold nanomaterials as HR-TEM calibration standard. Elsevier, Micron. 2015;**79**: 46-52

[37] Elastic strain engineering for unprecedented materials properties. In: MRS Bulletin. Li J, Shan Z, Ma E. Vol. 39 (2). 2014: p. 56. Available from: http://www.mrs.org/bulletin

[38] Goldsmid H. Porous thermoelectric materials. Materials. 2009;**2**:903-910. DOI: 10.3390/ma2030903

[39] Kuznetsov VL. Functionally Graded Materials for Thermoelectric Applications. In: Rowe DM (Ed.). Thermoelectrics Handbook: Macro to Nano (1st ed.). CRC Press; 2006. DOI: 10.1201/9781420038903

[40] Sweidan F, JinRyu H. One-step functionally graded materials fabrication using ultra-large temperature gradients obtained through finite element analysis of field-assisted sintering technique. Materials and Design. 2020;**192**:108714, Elsevier. DOI: 10.1016/j.matdes.2020.1087

[41] Yatsyshyn S, Stadnyk B, Melnyk K, Skoropad P, Baitsar R. Nanothermodynamics and nanometrology. In: Cyber-Physical Systems and Metrology 4.0. Barcelona: IFSA Publishing; 2021. pp. 105-157

[42] Ivanova A, Fuksov V, Gerasimov S, Pohodun A. Evolution of the inner liquid–solid interface during metal freezing. International Journal of Thermophysics. 2017;**38**:2. DOI: 10.1007/s10765-016-2168-4

[43] Bobalo Y, Kolodiy Z, Yatsyshyn S, Stadnyk B. Development of noise measurements. Part 3. Passive method of electronic elements quality characterization. Sensors and Transducers. 2013;**152**(5):164-168

[44] Stadnyk B, Yatsyshyn S, Ya. L, Datsiuk M. Development of noise measurements. Part 8. Nanometrology and nanothermodynamics as its scientific basis. Sensors and Transducers. 2013;**160**(12):25-34

[45] Stadnyk B, Yatsyshyn S, Seheda O. Research in nanothermometry. Part 6: Metrology of Raman thermometer with universal calibration artifacts. Sensors and Transducers. 2012;**142**(7):1-9

[46] Dreyfus PA, Psarommatis F, May G, Kiritsis D. Virtual metrology as an approach for product quality estimation in industry 4.0: A systematic review and integrative conceptual framework. International Journal of Production Research. 2022;**60**(2):742-776. DOI: 10.1080/00207543.2021.1976433

[47] Jun S, Kochan O, Kochan R. Thermocouples with built-in self-testing. International Journal of Thermophysics. 2016;**37**(4):1-9

[48] Kochan O, Kochan R, Bojko O, Chyrka M. Temperature measurement system based on thermocouple with controlled temperature field. In: Proceedings of Fourth IEEE Workshop on Intelligent Data Acquisition and Advanced Computing Systems: Technology and Applications. IEEE; 2007. pp. 47-50

Chapter 8

Thermodynamic Analysis of Thermoelectric Legs with Different Geometric Parameters

Alexander Vargas Almeida and Miguel Angel Olivares-Robles

Abstract

The design of thermocouple legs is the central theme of this chapter, the methodology of reduced variables is shown, which allows the designer to obtain the dimensions of the geometric parameters, specifically the cross-sectional areas transversal (A_n), (A_p) and length of the legs $(l_n = l_p)$. The main quantities used within this scheme calculation method are the reduced current density (U), the thermoelectric potential (Φ), and the thermal conductivity (κ). Subsequently, the performance of the designed thermocouple is analyzed through the electrical power produced when the system is connected to an electrical resistance load. In this step, the condition of the dependence of thermoelectric properties on temperature is used. The results show specific values of the cross-sectional areas in which the maximum power value occurs. Although the main method has the advantage that it uses minimal computing and software requirements, a spreadsheet may be sufficient for the calculations. It is also important to mention that in addition to calculating the dimensional parameters, the technique allows the generation and evaluation of various designs for the same temperature range.

Keywords: thermoelectric, thermodynamic, geometric parameters, thermoelectric generator, thermocouples

1. Introduction

The global energy demand is increasing every day [1] as a function of population growth. Most of the supply is achieved through fossil fuels (coal, natural gas, oil), [2]. These are distributed in different proportions over the planet, and it has an economic-environmental cost they imply for their exploitation. The projections made by various organizations indicate that energy dependence on fossil fuels will continue for many decades [2]. However, they also show encouraging results about better use of available energy resources (whether conventional or renewable) [3]. One form of this use is waste heat capture [1], which consists of implementing systems that capture residual heat from other generation systems and convert it into some form of useful energy. Thermoelectric generators have emerged as a response to this type of thermal use [1, 4].

A thermoelectric generator is a device capable of converting heat into electricity directly. Its basic structure [5, 6] is made up of a set of pairs of semiconductor materials (thermocouples), connected electrically in series and thermally in parallel. This entire arrangement is encapsulated between two rectangular plates of ceramic material. This device is linked on each side to a pair of heat exchangers, one of them being the hot side and the other the cold side. In this way, the module is under a temperature difference. When circulating a flow of heat Through the semiconductor materials (one type *N* and the other type *P*), the module produces a direct electric current, giving rise to a phenomenon known as the Seebeck effect.

Thermoelectric phenomena were discovered in the early nineteenth century, first by Thomas J. Seebeck [6, 7]. Later, Jean C. A. Peltier discovered the reverse phenomenon of cooling. The first applications of these two phenomena occurred during and after the Second World War, mainly for military use [8] with 5% efficiency for power generation. Advances continued towards domestic applications, developing some devices such as small radios that took advantage of the heat emitted by oil lamps. At the end of the 1960s, interest in thermoelectric technology declined, and many research programs in this field were dismantled. Despite this decline, Abram Fedorovich Ioffe and his research group at the institute in St. Petersburg carried out research on thermoelectrics in the USSR [9], the result of these efforts being the first thermoelectric generating and cooling devices. Around 1970, the reliability and simplicity of thermoelectric generators led them to consolidate among the technologies used for space exploration missions, such as the radioisotope thermoelectric generator used by NASA on the Apollo missions (1969–1972), Pioneer (1972–1973), Viking (1975), Voyager (1977), Galileo (1989), Cassini (1997), and Curiosity (2011) [8, 9]. Currently, the area of thermoelectricity involves both intense research and industry with a global impact because of the synergy between academia and industry. A variety of applications have been generated, including the automotive industry to harness the heat produced in automobile internal combustion engines, power supplies for electronic devices, self-powered wireless micro-platforms, and health monitoring systems [1].

Similarly, as various systems (mechanical, thermal, electrical, electronic) manufacture thermoelectric modules, it is necessary to elaborate the design of its structure, explicitly taking into account the influence of the physical phenomena that occur in the material network. For example, the design of thermocouple legs is carried out by taking advantage of the relationship between the transport mechanisms (of heat and electrical charge) and the geometric or dimensional parameters of the legs or thermoelectric pillars. Since the development of thermoelectric technology, thermocouples have been manufactured in a rectangular shape and with legs of the same cross-sectional area and length. However, with the development of new product concepts such as additive manufacturing, it has been possible to think of new geometric shapes for the legs.

For example, in [10] hollow prism geometries with square, triangular and circular bases are proposed, solid legs with circular, square, and triangular bases are also proposed, as well as legs of stacked layers with triangular, quadrangular, and circular shapes. Their results conclude that thermocouples with complex geometry (specifically stacked-layer legs) can achieve high electrical power values compared to legs of conventional shape. In [11] a study of optimization of the design of micro thermoelectric generators is shown, and it analyzes the role of structural geometry in the increase of the electrical power of the devices. One of its results shows that the cross-sectional area of the legs has a significant impact on the power and efficiency of the device.

The interest of the content of this chapter is to show that the design of thermoelectric legs is related to the needs of the application in which the thermoelectric generator will be used to harness heat. For example, in [12] it is shown that by varying the geometry of the leg of cascade-type thermoelectric modules, the performance of solar thermoelectric generators can be improved.

This work is focused on the design of thermocouple legs, presenting a first technique that is characterized by being of utility to the designer to carry out the calculations directly, preserving the physical information related to the phenomena of transport of electric charge and heat.

2. The thermoelectric generator as a heat engine

A heat engine is defined as a system that transforms heat into work in thermodynamics. It is a device that, operating cyclically, takes heat from a hot source, performs a certain amount of work (part of which is used to run the machine itself), and delivers waste heat to a cold source, usually the environment. The fluid known as the working substance must circulate through the structure of the machine and transports the heat from one source to another. In a broad sense, the term heat engine includes systems that produce work either through heat transfer or combustion, even though the devices do not operate in a thermodynamic cycle. According to what is mentioned in this last definition, it is possible then establish a thermoelectric generator as a heat engine, see **Figure 1**.

Due to the way it works, a thermoelectric generator is between two heat sources, one at high temperature or hot side temperature (T_H) and the other at low temperature or cold side temperature (T_C). Thus the *TEG* absorbs a quantity of heat, part of which is used to produce electrical work. It is important to clarify that unlike a traditional thermal engine, such as a steam turbine or an internal combustion engine, in the *TEG*, this role is played by the set of free electrons, which upon receiving the

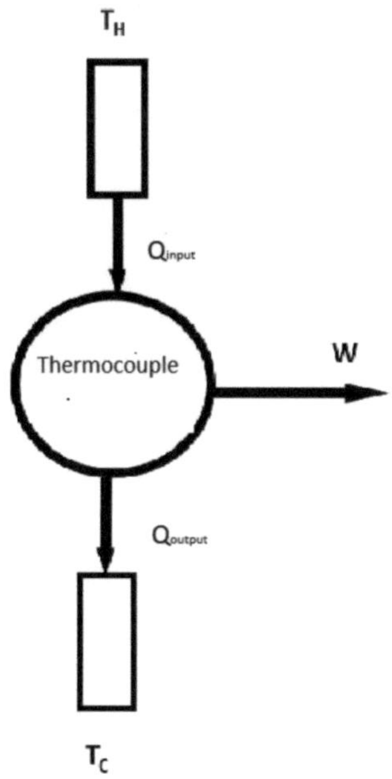

Figure 1.
Schematic to represent the thermoelectric generator as a heat engine.

energy transferred by heat, manage to produce an electric current through the Seebeck effect. If some load (R_{load}) is connected to the generator terminals, then a potential difference (ΔV) will be produced, and finally, an electrical work.

2.1 Thermal efficiency (η)

As is known in thermodynamics, thermal efficiency (η) measures the degree to which a thermal engine manages to take advantage of the heat that enters the hot side in the form of work. In terms of heat quantities, the thermal efficiency is formulated as follows,

$$\eta = \frac{Q_H - QC}{Q_H} = 1 - \frac{Q_C}{Q_H} \tag{1}$$

the theoretical limit of the thermal efficiency is imposed by the Carnot efficiency,

$$\eta_{max} = \eta_{Carnot} = 1 - \frac{T_C}{T_H} \tag{2}$$

the Carnot efficiency imposes the theoretical limit of the thermal efficiency,

$$\eta = \frac{\Delta T}{T_H} \frac{\sqrt{1 + Z\overline{T}} - 1}{\sqrt{1 + Z\overline{T}} + \frac{T_C}{T_H}} \tag{3}$$

where, ΔT is the temperature difference, $\overline{T}$ is the average temperature at which the device is operating, Z is the figure of merit and is given by,

$$Z = \frac{\alpha^2}{\rho\kappa} \tag{4}$$

it is observed that this quantity depends on the thermoelectric properties of the material, Seebeck coefficient (α), resistivity electrical ρ and thermal conductivity κ.

Finally, it is worth mentioning that the thermal efficiency of any machine is affected by different factors. Specifically for the *TEG*, it is the irreversible mechanisms of heat transport that negatively affect the system's performance. The following section deals with transport phenomena (reversible and irreversible) that occur in a thermoelectric generator.

3. Reversible and irreversible transport phenomena in thermocouples

In a thermoelectric material, two flows occur simultaneously, one of heat and the other of electric charge. There are three transport mechanisms for these flows in a thermoelectric generator: the Seebeck effect, the Joule effect, and the Fourier effect. **Table 1** shows the classification of these effects, depending on whether they are reversible or irreversible.

For **Table 1** the following variables are defined.

$\alpha \equiv$ Seebeck coefficient.

$T_H \equiv$ hot side temperature.

Effect	Formula	Classification
Seebeck	$\alpha T_H I$	*Reversible*
Joule	$\frac{1}{2}RI^2$	*Irreversible*
Fourier	$K(T_H - T_C)$	*Irreversible*

Table 1.
Mechanisms of heat and electric charge transport in a TEG.

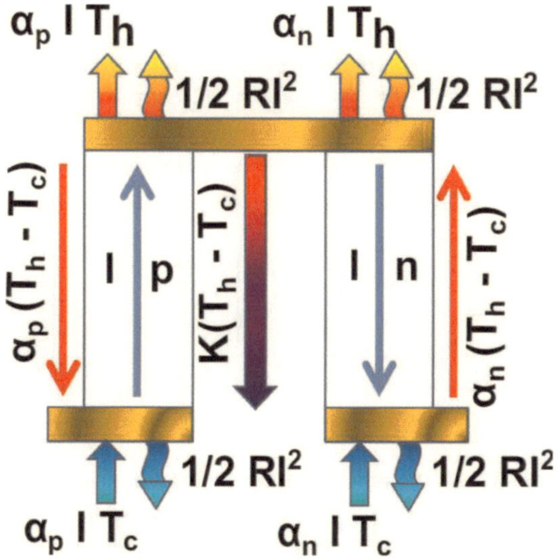

Figure 2.
Diagram showing the basic model of a thermocouple and the phenomena of heat and electric charge transport. The terms that are observed are the following: $\alpha_{n/p}(T_h - T_c)$ is the Seebeck voltage, $T_h - T_c$ is the temperature difference, α_{T_h/T_c} is the Peltier heat flow, $\frac{1}{2}RI^2$ is the heat dissipated by joule effect, $K(T_h - T_c)$ is the Fourier effect conduction heat.

$I \equiv$ electrical current.
$R \equiv$ electrical resistance.
$K \equiv$ thermal conductance.
$T_C \equiv$ cold side temperature.

The basic model is shown to show how these transport phenomena occur in a *TEG*, which is formed by two legs, one type n and another type p. This piece retains all the thermoelectric properties and maintains the dimensions of the original system, **Figure 2**.

The figure shows the direction in which each of the three transport mechanisms occurs, the Seebeck effect heat and the Fourier effect heat are due to the difference in temperature between the hot and cold sources. In contrast, the Joule effect heat is produced by the electrical resistance of the thermoelectric material itself.

4. The thermoelectric potential

As for other systems and processes that involve energy, thermoelectricity has sought to develop calculation schemes that are solid and that have the greatest possibility of being able to work with the information of the variables that correspond to the physics of the system. A treatment that has arisen from thermodynamic principles and that has shown a great scope demonstrated by its application to the design of thermocouples is the method of reduced variables, where one of the principal quantities is the thermoelectric potential (Φ) [13, 14]; this quantity is a state function given by the following equation,

$$\Phi = \frac{1}{u} + \alpha T \tag{5}$$

where, u is the reduced current density and α is the Seebeck coefficient.

A characteristic that shows the usefulness and importance of Φ is that it manages to link the two flows that occur simultaneously in the thermoelectric material of the leg, this quality becomes evident when deriving expressions for the volumetric production of heat (ν_q) and the electric field (E); as a function of thermoelectric potential Φ and current density J,

$$\nu_q = \nabla \cdot (\Phi J) = \nabla \Phi \cdot J \tag{6}$$

$$E = \nabla \Phi \tag{7}$$

The advantage of the thermoelectric potential (Φ) is that it allows working with classical thermodynamics, formulated from average quantities, even when the phenomenon has a microscopic nature and that the formalism that is applied for its treatment is the linear thermodynamics of coupled irreversible processes out of balance. Being formulated (Φ) as a state function, it is possible to analyze the system's evolution without taking into account the process path.

The following section shows the fundamentals of the reduced variables scheme, within which the formulation of the thermoelectric potential is generated as a result.

5. Reduced variables

When a system is analyzed in thermodynamics, a first step is to identify the variables that describe its behavior; subsequently, from this set, those that turn out to be the independent variables and whose numerical values can be obtained from an experiment or direct measurements are determined. Then, by applying the laws of thermodynamics, mathematical relationships between the dependent and independent variables are obtained, managing to deduce an equation of state. Although this equation can have a useful form to derive the quantities that control the thermodynamics of the system, in some instances, it is desirable to obtain a more straightforward form that allows a better understanding of the physics of the system. To achieve this goal, a useful technique is the application of reduced variables, which are intensive properties and thanks to which the equation of state can have a simpler form. However, it is also an equation that maintains the information of the process and allows analytical calculations of properties or design parameters in a practical and direct way. Snyder and collaborators [13] have built on solid foundations the method of reduced variables and have proposed a series of equations that allow calculations for the design of thermocouples.

For a complete understanding of the deduction of this method, it is suggested to review, for the scope of this chapter, it is sufficient to show and indicate the usefulness of the following equations. See **Table 2**.

Equation	Name	Interpretation
$u = \frac{J}{\kappa \nabla T}$	reduced current density	converts to total current density in an intensive variable
$\eta_r = \frac{u(\alpha - u\rho\kappa)}{u\alpha + \frac{1}{T}}$	reduced efficiency	a totally intensive quantity in terms of the reduced current density and of the temperature
$\eta_{r-max} = \frac{\sqrt{1+zT}-1}{\sqrt{1+zT}+1}$	reduced maximum efficiency	the highest value of the reduced efficiency
$s = \frac{\sqrt{1+zT}-1}{\alpha T}$	compatibility factor	the reduced current value that maximized to reduced efficiency

Table 2.
Main equations of the reduced variables method.

As can be seen in the **Table 2**, the new formulation of the efficiency η_r is found in terms of intensive quantities. This characteristic eliminates a considerable part of the interdependence between variables, and then the scheme becomes a practical tool for calculating thermocouple design. Specifically, this chapter shows its usefulness for designing conventional and segmented legs. The following section shows the application of this calculation scheme to design a conventional leg.

6. Geometric parameters of thermocouples

Figure 3 shows the basic structure of a thermocouple. In the diagram shown, it can be seen that the model is made up of a pair of legs of semiconductor material. Each of these legs has doping (one is n-type and the other p-type) and are connected (to each other) electrically in series employing a metal bridge.

Regarding system design, it is essential to say that this process can be as simple or complex as required for the application it seeks to serve. **Table 3** classifies the aspects that are taken into account to carry out the task of thermocouple design.

This chapter focuses on the design aspect of the size of the legs of the thermocouple. The geometric parameters of interest that will be calculated are the length (l) and cross-sectional area (A) of the legs and the thickness of the metal bridge. It is worth mentioning that this is the first scope and that the study can be extended and more complete when considering other parameters such as the thickness of the ceramic plate. In the following section, the design of a conventional thermocouple is carried out, taking into account the geometric parameters (l, A, l_{metal}).

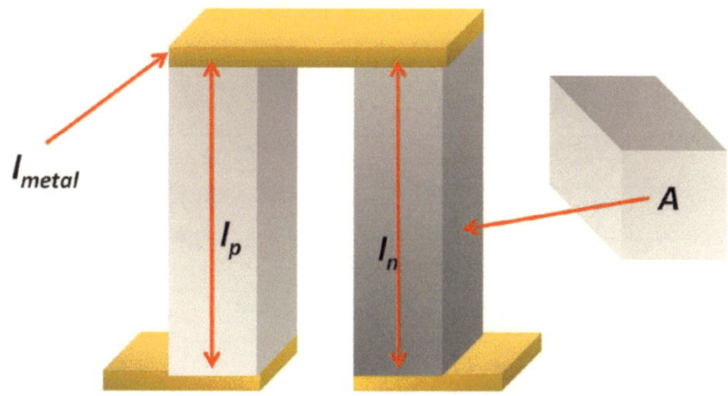

Figure 3.
Basic structure of a thermocouple.

Design aspect	Parameter type	Parameters
Composition	Material	Material of the metal bridge, material of the legs Ceramic plate material, amount of material to use
Size	Geometric parameters	Leg length, leg cross-sectional area, Ceramic plate thickness, metal bridge thickness
Architecture	shape	shape of the cross section of the leg, shape of the *TEG*, separation between the legs of the thermocouples

Table 3.
Classification of aspects and the corresponding parameters, for the design of thermocouples.

7. Modeling of conventional thermocouples

In **Figure 3** structure of a conventional thermocouple is shown. In this section, the design of a system is made to operate in the range of $(498K - 623K)$. For this purpose, the following parameters are calculated.

l is leg length.

A_n is cross-sectional area of n-type leg.

A_p is cross-sectional area of p-type leg.

l_{metal} is thickness of the metal bridge that electrically connects the two legs.

The materials selected for this first design are:

$p - type$: Zn_4Sb_3.

$n - type$: $CoSb_3$.

For the calculations, the equations of the **Table 2** are applied. Employing the definition of reduced current density, equation of the first row, the following integrals are obtained,

$$l(T)J_n = \int_{T_c}^{T_h} u_n \kappa_n dT \tag{8}$$

$$l(T)J_p = \int_{T_c}^{T_h} u_p \kappa_p dT \tag{9}$$

for the calculation of the integrals (8 and 9) the data of **Table 4** are necessary.

T(K)	$u_p k_p dT(A/cm)$	$u_n k_n dT$ (A/cm)
$T_0 = 498$	$u_p k_p(T_0) = 0 \cdot 4999$	$u_n k_n(T_0) = -2 \cdot 4164$
$T_1 = 523$	$u_p k_p(T_1) = 0 \cdot 5107$	$u_n k_n(T_1) = -2 \cdot 3984$
$T_2 = 548$	$u_p k_p(T_2) = 0 \cdot 5257$	$u_n k_n(T_2) = -2 \cdot 3855$
$T_3 = 573$	$u_p k_p(T_3) = 0 \cdot 5430$	$u_n k_n(T_3) = -2 \cdot 3783$
$T_4 = 598$	$u_p k_p(T_4) = 0 \cdot 5607$	$u_n k_n(T_4) = -2 \cdot 3774$
$T_5 = 623$	$u_p k_p(T_5) = 0 \cdot 5768$	$u_n k_n(T_5) = -2 \cdot 3832$
Source [14].		

Table 4.
Numerical data of the product uk for the p-type and n-type materials, in the temperature range $498 - 623K$.

Combining the data in **Table 4** with the fourth-order Newton-Cotes numerical method to calculate the integrals (8 and 9),

$$l(T)\int_{T_c}^{T_h} u_p \kappa_p dT = 52.7163\, mA/cm \tag{10}$$

$$l(T)\int_{T_c}^{T_h} u_n \kappa_n dT = -238.93\, mA/cm \tag{11}$$

then the quotient is calculated.

$$\frac{-J_n}{J_p} = 4.53237 \tag{12}$$

Next, it is to know the values of the total heat flux (W) and the thermoelectric potential (Φ) at the hot spot (T_h) of the thermocouple for each of the two legs (n) and (p). It is important to mention that the value of (W) is an input parameter that is obtained depending on the heat flow that is required for the application or final use of the *TEG*; the value of (Φ) is calculated with Eq. (5). For the particular case of the thermocouple designed, the values used for the design are given by the following **Table 5**.

The following calculation is that of the current densities $\left(J_p\right)$ and (J_n), applying the following eq. (13), $\left(J_p\right)$ is calculated.

$$J_p = W\frac{1 + \frac{A_n}{A_p}}{\Phi_p - \Phi_n} \tag{13}$$

to calculate (J_n) the definition of the quotient (14) is applied

$$\frac{-J_n}{J_p} \tag{14}$$

then the following values are obtained from the **Table 6**.

Parameter	Numerical value
W	20 (W/cm^2)
Φ_p	0.39643 (V)
Φ_n	−0.50663 (V)

Table 5.
Numerical values of thermoelectric potentials and heat flux, used for calculation of current density J_p.

Current density	Numerical value
J_p	27.0333 (mA/cm^2)
J_n	122.525 (mA/cm^2)

Table 6.
Calculated values of the current densities, J_p and J_n.

7.1 Calculation of geometric parameters

With the results obtained in the previous section, it is now possible to determine the numerical values of the geometric parameters. First, the length (l) of the thermocouples is calculated.

$$l = \frac{\int_{T_c}^{T_h} u_p \kappa_p dT}{J_p} \tag{15}$$

combining the result of (10) and (J_p) from **Table 6**.

$$l = 1.95\,mm \tag{16}$$

The following parameter to calculate is the cross-sectional area, A_t is defined as the cross-sectional area,

$$A_t = A_p + A_n$$

For a first calculation, a value $A_t = 1mm^2$ is proposed later to calculate A_p and A_n, the following system of linear equations is constructed,

$$A_p = 4.53A_n \tag{17}$$

$$A_p + A_n = 1 \tag{18}$$

When solving the system, the following values are obtained,

$$A_p = 0.82mm^2 \tag{19}$$

$$A_n = 0.18mm^2 \tag{20}$$

7.2 Thermoelectric properties of the thermocouple

This section is continued with the design of the thermocouple. The next step is to carry out part of the composition design (first row of **Table 3**) for the scope of this chapter, which is to show a first approach to the design of thermocouples, an of the most used techniques and with a certain degree of reliability, which is the calculation of the average. The formula for calculating the average of a thermoelectric property (x) in a temperature range ($T_c - T_h$) is defined as,

$$\overline{X} = \frac{\int_{T_c}^{T_h} X dT}{T_h - T_c} \tag{21}$$

Eq. (21) is applied to calculate the averaged properties: Average Seebeck coefficient ($\overline{\alpha}$), thermal conductivity average ($\overline{\kappa}$), average electrical resistivity (ρ). To complete this task, it is necessary to know the measurement data of the properties mentioned above. The following segment shows the calculation of the averaged amounts.

7.2.1 Average Seebeck coefficient

Measurement data for this quantity is shown in the **Tables 7–12** for the materials Zn_4Sb_3, $CoSb_3$;

The results of the averages are as follows.

$\overline{\alpha}_p = 177.56\mu V/K.$

$\overline{\alpha}_n = -178.84\mu V/K.$

Temperature (K)	Seebeck coefficient μV/K (Zn_4Sb_3)
498	168
523	173
548	178
573	182
598	187

From reference [14].

Table 7.
Numerical data of Seebeck coefficient of p-type material.

Temperature (K)	Seebeck coefficient μ V/K ($CoSb_3$)
498	−168
523	−174
548	−179
573	−184
598	−188

From reference [14].

Table 8.
Numerical data of Seebeck coefficient of n-type material.

Temperature (K)	Electrical resistivity Ωcm (Zn_4Sb_3)
498	2.691
523	2.760
548	2.825
573	2.889
598	2.949

From reference [14].

Table 9.
Numerical data of electrical resistivity of p-type material.

Temperature (K)	Electrical resistivity Ωcm ($CoSb_3$)
498	0.845
523	0.867
548	0.888
573	0.907
598	0.924

From reference [14].

Table 10.
Numerical data of electrical resistivity of n-type material.

Temperature (K)	Thermal conductivity (m W/cm K) (Zn_4Sb_3)
498	5.91
523	5.93
548	6.00
573	6.10
598	6.22

*From reference [14] (**Table 12**).*

Table 11.
Numerical data of thermal conductivity of p-type material.

Temperature (K)	Thermal conductivity (m W/cm K) ($CoSb_3$)
498	39.925
523	39.10
548	38.37
573	37.75
598	37.25

From reference [14].

Table 12.
Numerical data of thermal conductivity of n-type material.

7.2.2 Average electrical resistivity

Measurement data for this quantity is shown in the **Tables 9** and **10** for the materials Zn_4Sb_3, $CoSb_3$;

the results of the averages are as follows.

$\overline{\rho}_p = 2.82\Omega cm.$

$\overline{\rho}_n = 0.89\Omega cm.$

7.2.3 Average thermal conductivity

Measurement data for this quantity is shown in the Tables for the materials Zn_4Sb_3, $CoSb_3$; (**Table 11**).

the results of the averages are as follows.

$\overline{\kappa}_p = 6.02 m\, W/cm K.$

$\overline{\kappa}_n = 38.44 m\, W/cm K.$

8. The metallic bridge and its properties

The connection between two legs, one type (n) and the other type (p), to form a thermocouple is achieved by making a bridge-type connection between a pair of ends (one of each leg) through a segment of metallic material, as shown in **Figure 4**.

From the figure, it is possible to understand that the metallic material is responsible for transporting charge carriers from one terminal to another when the thermocouple is under heat transport. So the designer must consider the effect of this part on the structure of a thermoelectric system. **Table 13** shows the numerical values of the thermoelectric properties of the metallic bridge,

9. Final design

The previous sections have shown the first thermocouple design procedure in the most detailed way possible. **Figure 5** shows a sketch of the design generated so far,

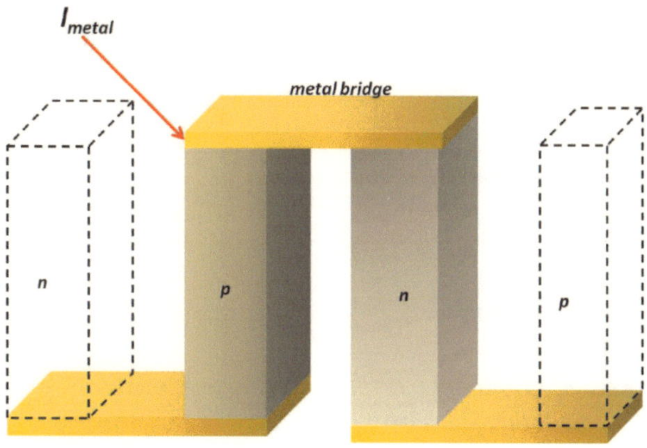

Figure 4.
Connection between the legs of the thermocouple.

Component	Electrical resistivity (Ωcm)	Thermal conductivity (W/cm K)
Metal	$1 \cdot 7 * 10^{-6}$	$3 \cdot 98$

Data obtained from [15, 16].

Table 13.
Thermoelectric properties of the metal bridge.

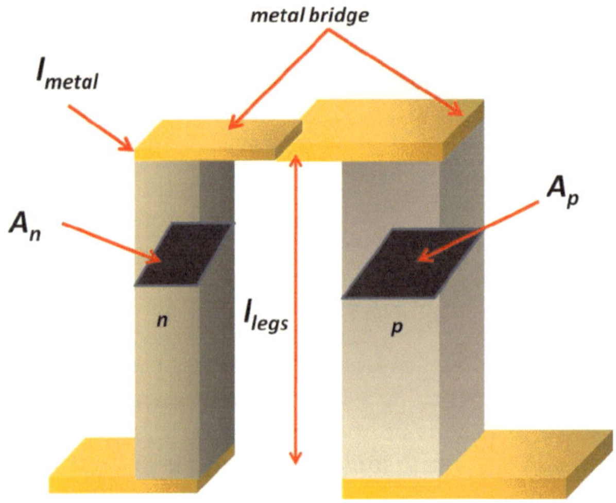

Figure 5.
Thermocouple structure designed. The dimensioning of the length of the legs, the cross-sectional area of each one and the thickness of the metal bridge are shown.

Characteristic	Symbol	Numerical value
Average Seebeck coefficient of p-type material	$\overline{\alpha}_p$	177.56 μ V/K
Average Seebeck coefficient of the n-type material	$\overline{\alpha}_n$	- 178.84 μ V/K
Average electrical resistivity of p-type material	$\overline{\rho}_p$	2.82 Ω cm
Average electrical resistivity of n-type material	$\overline{\rho}_n$	0.89 Ω cm
Average thermal conductivity of p-type material	$\overline{\kappa}_p$	6.02 m W/cm K
Average thermal conductivity of n-type material	$\overline{\kappa}_n$	38.44 m W/cm K
Cross-sectional area of p-type leg	A_p	0.82 mm^2
Cross-sectional area of the n-type leg	A_n	0.18 mm^2
Leg length	l	1.95 mm

Table 14.
Characteristics of the designed thermocouple.

The system obtained is characterized by having legs of the same length but of different cross-sectional areas. It is important to note that the cross-sectional areas remain constant throughout the length of each of the legs. The legs are joined (one end of each) by the metal bridge, which has an area that is adjusted to each of the cross-sectional areas (A_p, A_n). **Table 14** shows a file of the design characteristics,

The designed system is already characterized. The next step is to analyze its performance. In the next section, it is performed this analysis considers certain conditions regarding the geometric parameters.

10. Analysis of the performance of the designed thermocouple

The performance of the designed thermocouple is evaluated using the power (P), which has the following equation,

Symbol	Electrical resistance type
R_p	Electrical resistance of the p-type leg
R_n	Electrical resistance of the n-type leg
R_{metal}	Electrical resistance of the metal bridge
R_{load}	System load resistance connected to the thermocouple

Table 15.
Types of electrical resistance.

$$P = VI \tag{22}$$

where,
P: electrical power generated by the thermocouple, (W),
V: voltage produced by the thermocouple, $(Volts)$,
I: current produced by the thermocouple; $(Amperes)$.
The corresponding formulation for the quantities (V), (I) is as follows:

$$V = \frac{(\alpha_p + \alpha_n) R_{load}}{R_n + R_p + R_{metal} + R_{load}} (T_H - T_C) \tag{23}$$

$$I = \frac{(\alpha_p + \alpha_n)}{R_n + R_p + R_{metal} + R_{load}} (T_H - T_C) \tag{24}$$

in eqs. (23 and 24) the following electrical resistances are found, (**Table 15**).

Eqs. (23 and 24) are combined with eq. (22) to formulate the electrical power of the thermocouple. In the following subsections show specific cases with imposed conditions for the parameters.

10.1 Electrical power with variation of total cross-sectional area and electrical load resistance

In this first case, it is intended to observe what happens when the thermocouple is connected to a system that is fed with the current produced, as shown in **Figure 6**.

In the design calculations, carried out in the previous sections, a total area value was established $(A_{total} = 1mm^2)$, and determining the corresponding values of (A_p) and (A_n). Graph 7 shows how the power generated by the thermocouple varies with respect to the ranges $(0.1 \le A_n \le 0.18)$ and $(0.1 \le A_p \le 0.82)$ for a fixed value of resistance load R_{load}.

It is observed in the **Figure 7** that at the beginning of each of the ranges, the power surface grows rapidly and stabilizes when it reaches a maximum value. **Table 16** shows the maximum power value (P_{max}) and the values of (A_n) and (A_p) that allow the thermocouple to reach this maximum value.

It is important to note that the above results have been calculated for a specific value of load resistance (R_{load}). Below are two graphs that show the effect of the variation of (R_{load}) on the power.

The previous **Figures 8** and **9** show that as the load resistance increases, the electrical power produced decreases. **Figure 8** has been obtained for three different values of R_{load} in the following order: (5Ω) red color, (10Ω) blue color, (15Ω) orange color.

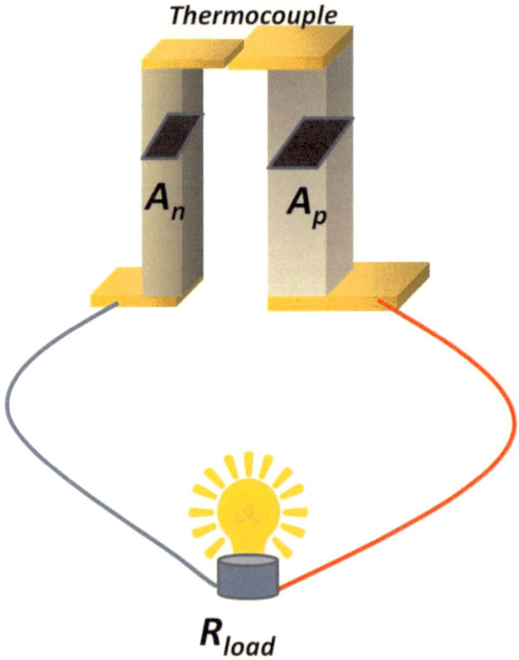

Figure 6.
Thermocouple connected to a resistance electrical load.

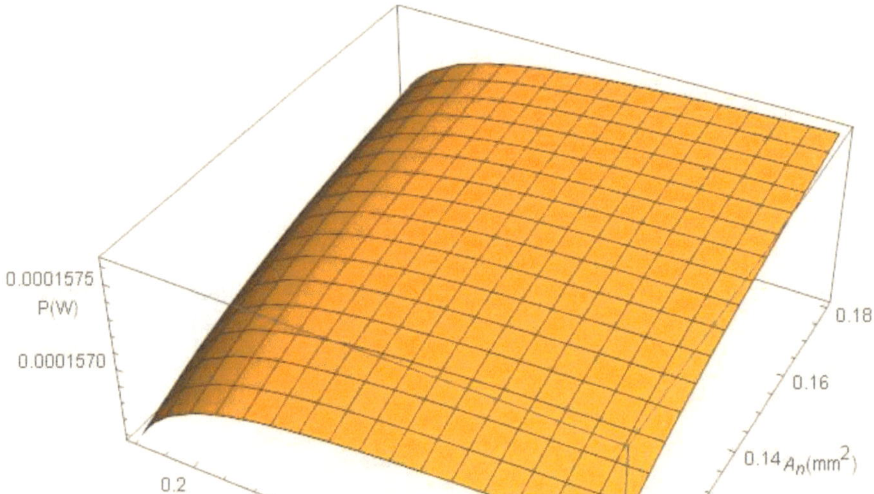

Figure 7.
Power surface produced by the thermocouple, as a function of cross-sectional area.

Parameter	Numerical value
P_{max}	0.00016
A_p	0.67
A_{total}	0.14

Table 16.
Maximum power value and corresponding cross-sectional area values.

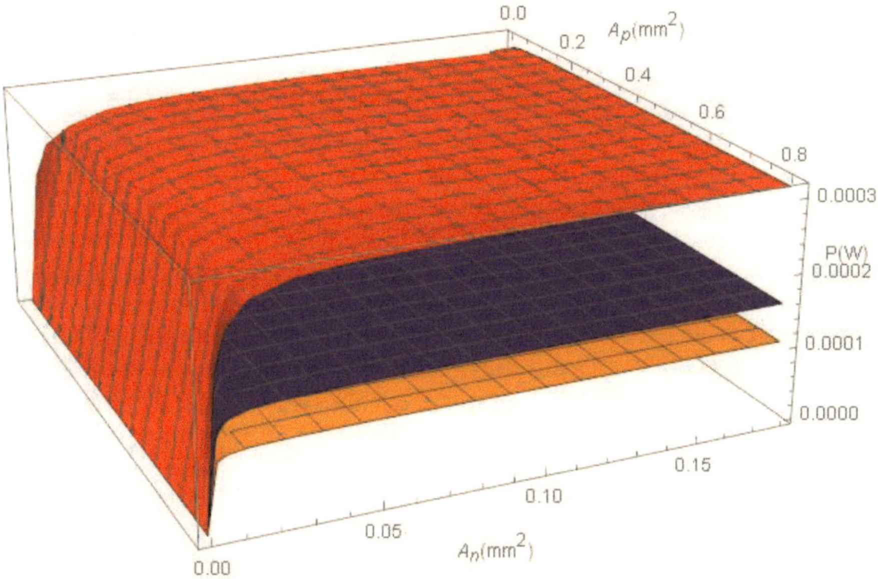

Figure 8.
Power surfaces for different values of electrical load resistance.

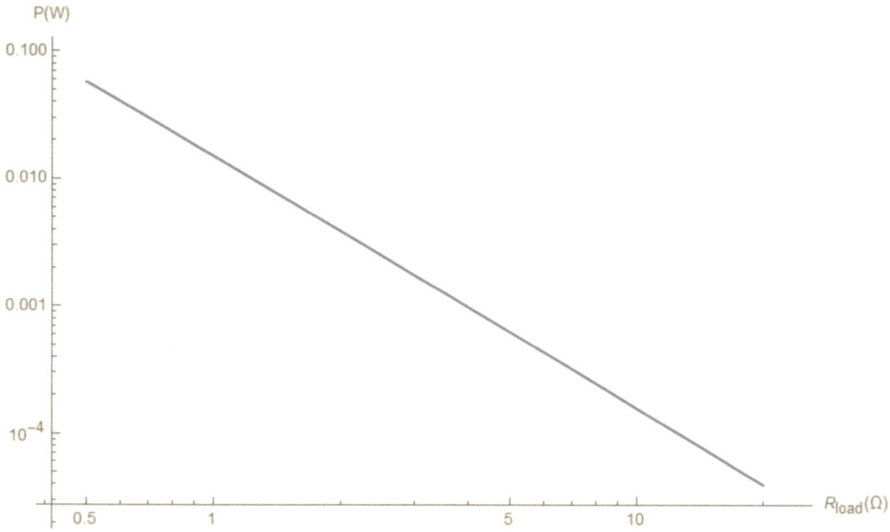

Figure 9.
Electrical power for fixed values of Ap and an, with variation of the electrical load resistance.

11. CPU time and compute resource characteristics

One type of information that is important to document when developing codes for numerical calculations is the characteristics of the computer equipment and the execution time of the codes. The design calculations were made using a personal computer, which has the characteristics shown in **Table 17**.

Figure 10 show the CPU time required for the execution of the calculation code of eqs. (10 and 11).

12. Conclusions

This chapter has shown a methodology for the design of thermocouple legs. This methodology is based on the approach of reduced variables. The reduced current density (u) is the main quantity combined with the thermoelectric potential (Φ) and the thermal conductivity. Then it is possible to calculate the length of the legs (l). Subsequently, setting a fixed value for the total cross-sectional area (An + Ap) and again using the quantities from the reduced variables approach has been possible to calculate the cross-sectional areas corresponding to each of the legs (A_n) and (A_p). It is important to note that numerical values of the thermoelectric properties (α, ρ, κ) of each of the two materials (Zn_4Sb_3) y $(CoSb_3)$ were also used to perform the calculations. In order to have support for the classification of thermocouple design tasks, a **Table 3** has been prepared; with the help of this resource, the designer can label the type of design being carried out.

CPU	AMD Ryzen 33,200 U with Radeon Vega Mobile Gfx 2.60 GHz
Installed RAM	8.00 GB (5.95 GB usable)
Type of system	64-bit operating system, x64 processor

Table 17.
Characteristics of the computer equipment used for design calculations.

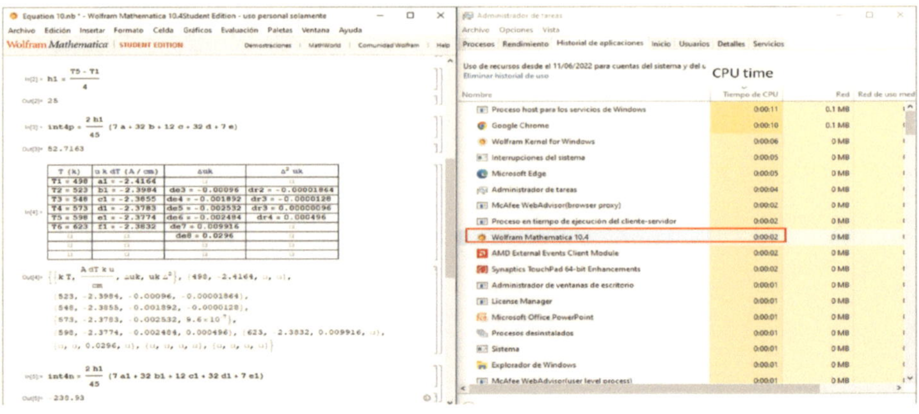

Figure 10.
CPU time required for the execution of the calculation code of eqs. (10 and 12).

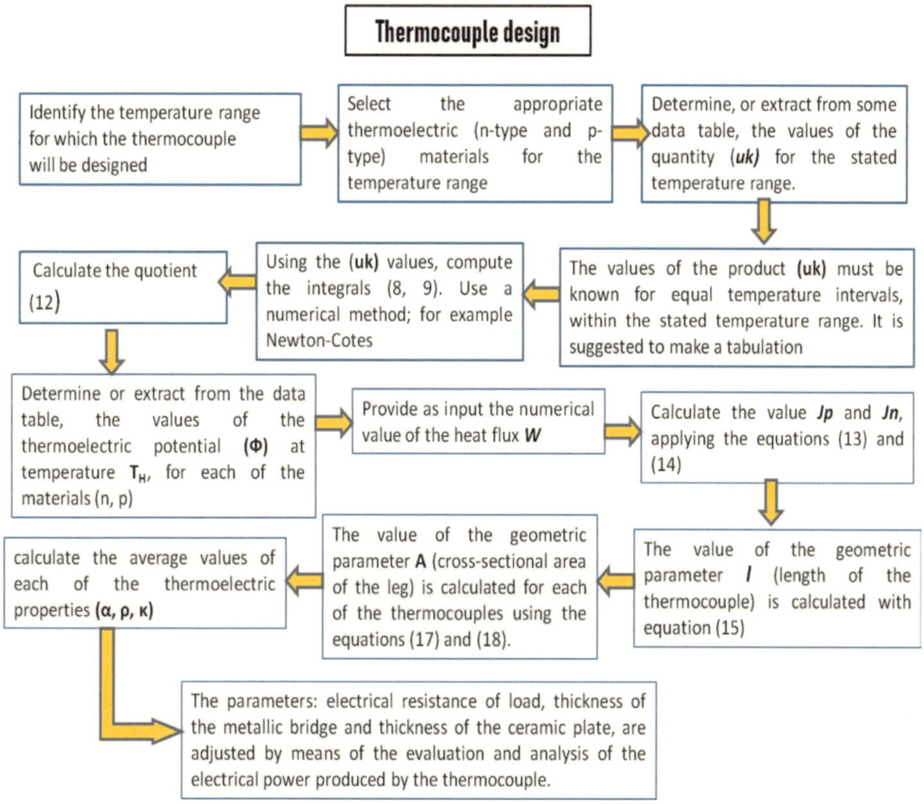

Figure 11.
Flowchart with the steps to design the thermocouple.

The methodology is also characterized by considering the dependence of thermoelectric properties on temperature. This aspect allows generating a specific design for a temperature range according to the application required to be served. In the example of the designed thermocouple, notice that the design of the metal bridge was included. In a later work, it is intended to add the sizing of the ceramic plate. The analysis of the power of the thermocouple shows that there are specific values of (A_n) and (A_p) that allow it to reach its maximum value. That is, it is possible to determine optimal values of the cross-sectional area of the legs for a particular load resistance (R_{load}). Graphs 8 and 9 lead to the conclusion that (R_{load}) also determines the maximum power value that the designed system can reach. The **Figure 11** shows the steps of the methodology that has been shown in this work for the design of thermocouples.

Acknowledgements

This research was funded by Instituto Politecnico Nacional, Mexico grant number 20220099 and CONACYT-Mexico grant number CVU 444915. The APC was funded by Instituto Politecnico Nacional, Mexico.

The authors acknowledge the editorial assistance in improving the manuscript.

Author details

Alexander Vargas Almeida[1] and Miguel Angel Olivares-Robles[2]*

1 Universidad Politécnica del Golfo de Mexico, Carretera Federal Malpaso - El Bellote, Monte Adentro, Paraíso, Tabasco

2 Instituto Politecnico Nacional, SEPI-ESIME-Culhuacan, Coyoacan, Ciudad de Mexico, Mexico

*Address all correspondence to: olivares@ipn.mx

References

[1] Jaziri N, Boughamoura A, Müller J. A comprehensive review of thermoelectric generators: Technologies and common applications. Elsevier – Energy Reports. 2020;**6**:264-287. DOI: 10.1016/j.egyr.2019.12.011

[2] International Energy Agency, World Energy Outlook. Available from: http://oilproduction.net/informes-reportes/item/312-world-energy-outlook-2010-resumen-ejecutivo

[3] Fernández Ramos A. Energy Harvesting: una técnica para generar energa [Tesis inédita de licenciatura]. Madrid, Spain: Universidad Autónoma de Madrid; 2017

[4] Pham HN. Design and optimization of effective segmented thermoelectric generator for waste heat recovery. Dinamarca: DTU library; 2015

[5] He J, Tritt TM. Advances in thermoelectric materials research: Looking back and moving forward. Science. 2017;**357**(1369):29. DOI: 10.1126/science.aak9997

[6] Ravindra NM, Jariwala B, Banobre A, Maske AA. Thermoelectrics: Fundamentals, Materials Selection, Properties, and Performance. Suiza: Springer; 2019

[7] Goupil C, Ouerdane H, Zabrocki K, Seifertt W, Hinsche NF, Müller E. Thermodynamics and Thermoelectricity en Continuum Theory and Modeling of Thermoelectric Elements. USA: Wiley; 2015. pp. 1-74

[8] Ursell TS, Snyder GJ. Compatibility of segmented thermoelectric generators. In: Twenty-First International Conference on Thermoelectrics, 2002. Proceedings ICT 02. Long Beach, CA, USA: IEEE; 2002. pp. 412-417. DOI: 10.1109/ICT.2002.1190349

[9] Brief History of Thermoelectrics. En Northwestern Materials Science and Engineering. Available from: http://thermoelectrics.matsci.northwestern.edu/thermoelectrics/history.html

[10] Dunham MT, Barako MT, LeBlanc S, Asheghi M, Chen B, Goodson KE. Power density optimization for micro thermoelectric generators. Energy. 2015;**93**(2):2006-2017. DOI: 10.1016/j.energy.2015.10.032

[11] Tanwar A, Lal S, Razeeb KM. Structural design optimization of micro-thermoelectric generator for wearable biomedical devices. Energies. 2021;**14**:2339. DOI: 10.3390/en14082339

[12] Maduabuchi C, Ejenakevwe K, Jacobs I, Ndukwe A, Mgbemene C. Analysis of a two-stage variable leg geometry solar thermoelectric generator. In: Proceedings of the 2nd African International Conference on Industrial Engineering and Operations Management Harare. Zimbabwe: IEOM Society International. December 7–10, 2020. 2020

[13] Snyder GJ, Ursell T. Thermoelectric efficiency and compatibility. Physical Review Letters. 2003;**91**:148301

[14] Snyder GJ. Thermoelectric power generation: Efficiency and compatibility. In: Thermoelectric Handbook: Macro to Nano 9–1–9-26. Boca Raton: CRC Press; 2006

[15] Available from: https://www.electronics-notes.com, Electrical Resistivity Table for Common Materials. Table of the electrical resistivity for materials that

may be used in electrical and electronic components, including the resistivity of copper, resistivity of brass and the resistivity of aluminium

[16] Available from: https://thermtest.com/thermal-resources/top-10-resources/top-10-thermally-conductive-materials. Top 10 Thermally Conductive Materials

Chapter 9

New Thermoelectric Future and It's Uses towards Mankind: A Review

Hiranmoy Samanta, Kamal Golui and Soumyadeep Mukherjee

Abstract

In the last few decades, the attention is being carried by the research and development of wearable sensors for the potential, optimization and hand ready data in instantaneous and reliable health monitoring for assessing the health of a person and default measures are taken care of in time. The idea of body heat based thermoelectric power generation permits an attractive solution which is used for thermoelectric power for wearable devices. This review article represents the different types of thermoelectric generators and the successive results which have been achieved till date. The paper also reflects the problems concerning the operation as well as the O/P of wearable sensors based on body heat harvesting method power generation. Specifically, the paper focuses on optimized simulation of human thermoregulatory models, flexible heat sinks, electronics, and energy storage devices. Which are pertinent in nature due to the application and alongside research which leads to the practical implementation of these sensors in practice for a better health monitoring and healthy lifestyle.

Keywords: thermoelectric material, generator, modulator, body heat harvesting, health monitoring

1. Introduction

Developing issues of the consumption of power resources indispensable to stylish life, along with oil, petroleum gas, and coal, are utilizing the improvement of late innovation fundamentally founded absolutely on the utilization of chance home grown resources: sun power, hydroelectric power, wind power, bioenergy, geothermal power, and so on. Taking everything into account, it possesses a one of a kind districts in human exercises, since it goes within the limits of the business strategies and techniques that emerge in nature. In most extreme cases, squander heat is lost without a money related benefit. These power helpful asset expenses cost nothing and might be utilized to diminish the impact of the power debacle and overall warming. Thus, the transformation of waste heat into energy [1–3] is to be invited. The heat-to-energy converter is known as thermoelectric generator (TEG) or thermoelectric stack (within the nineteenth century) [4]. The practical precept of the TEG is primarily based totally at the Seebeck impact found in 1821. The present day TEG [5] is a stack such as a mess of pairs A and B of various substances linked in collection through electric conductors.

The temperature distinction among facets of the TEG [5] causes every pair to generate an electrical potential, and the sum of those potentials is known as the electromotive pressure of the cell. By growing the range of those pairs, the electromotive pressure may be elevated to the favored value. The electromotive pressure of the stack will increase whilst the electric conductivity of the substances A and B is of specific nature, of type "n" (negative) and "p" (positive). Power turbines primarily based totally at the Seebeck impact are impartial of the kind of consumable warmth and might consequently be utilized in specific areas. It is essential to observe that the tool may be used now no longer handiest to transform heat to energy, however additionally the opposite process. When a cutting-edge of is fed to this apparatus, the temperature distinction among its facets is created (Peltier impact, found in 1834). In this case, the tool is known as Thermoelectric Cooler (TEC). TECs have been evolved within side the shape of Peltier-Peltier pellets for small-capability and space-restrained applications. They are frequently used to mood and funky digital components.

In this paper we have presented a brief literature review of the major works carried out so far in this field. Next thermoelectric materials have been described. Its working principle, efficiency measurement and its classification has been presented henceforth. How thermoelectric energy can be harvested using body heat [6] has been discussed in brief. A brief discussion on the working and functioning of thermoelectric generators and thermoelectric modules have been presented next. Applications of thermoelectric generator in different fields have been discussed too.

2. Literature review

1. Narducci and Giulio [7] presented and discussed advancement in research on silicon and related materials for thermoelectric applications, mostly focusing on the comparison between the two strategies deployed to increase its performance, namely either reducing its thermal conductivity or, in polycrystalline materials, increasing its power factor.

2. Zoui et al. [8] reviewed state-of-the-art thermoelectric generators, applications and recent progress in the field. A comparative study has been done by groping the applications of thermoelectricity into three main domains.

3. Nozariasbmarz et al. [9] demonstrated the performance of thermoelectric generators having low and high fill factor and compared in harvesting body heat.

4. Levinsky [10] has discussed the thermoelectric properties of different materials and their experimental methods as well. He gave brief information about all types of categories these materials could be divided into according to their possession of thermoelectric properties.

5. Mahan [11] has provided a brief introduction on the development of thermoelectric materials and its working. He studied, compared and validated the Seebeck coefficient of silicon and studied its utility as a thermoelectric material.

6. Polozine et al. [12] have discussed the development of thermoelectric materials for electricity generation. They rectified the criteria of Thermoelectric figure of merit and used it in their study and named it criterion of usefulness.

7. Leonov [13] studied harvesting of thermoelectric energy on people and found that although power generation is affected by many factors such as ambient temperature, wind speed, clothing thermal insulation, and a person's activity, it does not directly depend on metabolic rate. An experimental study has been carried out and found that a thermoelectric shirt with an energy harvester produces more energy during the same period of time than the energy stored in alkaline batteries of the same thickness and weight.

8. Vining [14] has put forward a truth about the thermoelectric materials that states that thermoelectric energy conversion will never be as efficient as steam engines. He also discussed its impact on the recent climate crisis.

9. Martin [15] has discussed the working of thermoelectric materials and their application. He compared the thermoelectric properties of different materials and showed the relationship between power factor, electrical conductivity, Seebeck coefficient and carrier concentration.

10. Riffat and Ma [16] have given basic knowledge of the thermoelectric devices and an overview of these applications are given. The prospects of the applications of the thermoelectric devices are also discussed.

11. Snyder [17] discussed the designing principles for thermoelectric generators in energy harvesting applications, and the various thermoelectric generators available or in development.

12. Kasap [18] has discussed the thermoelectric properties in a metal and studied its behavior using the Seebeck effect. He also discussed thermocouple and its principles.

3. Thermoelectric material

Thermoelectric devices (thermoelectric modules) can convert electrical energy into a temperature gradient, a phenomenon discovered by Peltier in 1834. Applications for this cooling or heating [19, 20] effect remain kept to a minimum until the drive semiconductor material is developed. With the advent of semiconductor materials has enabled many practical applications of thermoelectric devices.

The temperature distinction among factors in a conductor or semiconductor consequences in an ability distinction among those factors. In different words, a temperature gradient in a conductor or semiconductor offers upward push to an embedded electric powered field. This phenomenon is known as the Seebeck effect or the thermoelectric effect (**Figure 1**).

Thermoelectric materials are getting increasingly important as an alternate energy [1, 2] source, and therefore the applications [21] for these materials are increasing. The thermoelectric effect involves the generation of electricity from a heat source or the removal of warmth when an electrical current is used as a cloth. Thermoelectric devices are used for power generation and cooling [19, 20]. They're developed to get electricity from waste heat sources (cars, trucks, industrial processes, chemical processes, steel industries for example), space energy, remote low voltage power sources, human pellet coolers, car coolers, electronic coolers [1, 2] and refrigeration [22]. The technology is predicated on the Seebeck effect, which involves a voltage (power) generated during a material when a temperature difference is applied across it (heat flow) [21].

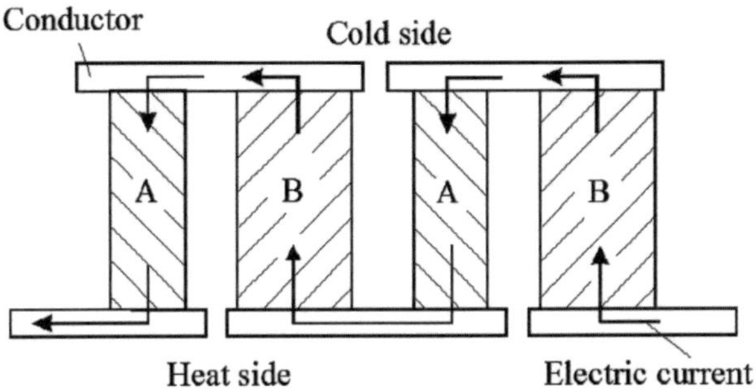

Figure 1.
Thermoelectric pile [12].

Thermoelectric materials have the ability to convert heat flows into electrical energy(Seebeck Effect) and vice versa (Peltier Effect). Their use is becoming more and more interesting, because they see the benefits of waste energy recycling [1–3]. It's about the transformation of heat from industry or trucking into electricity, thereby increasing the efficiency of the system and reducing operating costs and environmental pollution. Thermoelectric gadgets are especially solid, quiet and do not produce vibrations in light of the fact that their activity does not need contribution of mechanical energy. Thus, extensive endeavors have been made, utilizing new materials [23], to foster thermoelectric framework innovation. Since the disclosure of thermoelectricity (TE) in 1821 by Seebeck, specialists have attempted to comprehend and control this peculiarity. Peltier did it in 1834 when he found the contrary impact, and Lord Calvin in 1851 planned the law connecting the two peculiarities. In the next 100 years, in 1909, Edmund Altenkirch precisely determined the energy proficiency of a thermoelectric generator interestingly. In 1950, Abram Ioffe found the thermoelectric properties of semiconductors, which opened up new thermoelectricity projections with a worth near 1 [24]. This worth is still low, yet satisfactory enough for certain creators and industrialists to plan new applications to advertise. One such application is the thermoelectric fridge planned by Becket et al. in 1956. Around the same time, the possibility of thermoelectric generators [25] showed up, for example, Ioffé's 1957 thermoelectric light [24], which controlled radios by recuperating the intensity delivered by the light.

4. Measurement of efficiency of thermoelectric materials

Soon after the foundations of thermoelectric materials were laid down by Peltier and Seebeck, a widespread extensive research began for the development of various such life impactful devices. However there came a situation where characterization of these materials where required based on how efficiently they can convert heat to electricity or to ensure minimal loss of energy during operation. The figure of merit (zT) a numerical parameter was introduced which suggested that thermoelectric materials which obtained a figure of merit closer to unity was considered to be highly efficient and those with figure of merit less than unity were comparatively less efficient. However modern scientific methods have made it possible to synthesize thermoelectric materials with figure of merit greater than unity. This efficiency can be increased by a variety of methods, one of them is by introducing nanostructures which have shown significant increase in efficiency of thermoelectric generator (TEG) [5, 26] and semiconductors (**Figure 2**).

DOI: http://dx.doi.org/10.5772/intechopen.107954

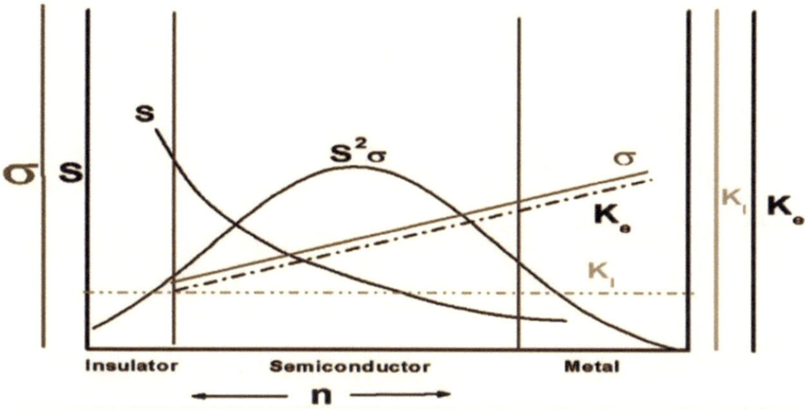

Figure 2.
Relationship between figure of merit ZT and other parameters such as electrical conductivity σ, Seebeck coefficient S, power factor $S^2\sigma$, electronic thermal conductivity Ke, thermal conductivity of the network Kl and total thermal conductivity K [27].

5. Conventional materials

These are semiconductor alloys or chemical compounds constituting an anion of the chalcogen family of the periodic table (group 16) and an electropositive element. These materials work between temperature ranges at which their efficiency is optimum. Bi-Te based materials work optimally at a temperature of less than 150°C, while materials based on Si-Ge are used at higher temperatures (around 500°C). Combination and segmenting of different materials can help us bring variance in these temperature ranges. The Bi2Te3 are notable and can have a figure of legitimacy near the unit at room temperature. In any case, as they are effectively oxidized and disintegrated, these materials can't be utilized for high-temperature applications in air. Around 70% of the TE modules available use Bismuth and Telluride as practical materials. Lead telluride (PbTe) is a decent thermoelectric material for applications requiring mid-temperatures up to 900 K. Melting point of this material being 1190 K it has great synthetic solidness, low fume pressure and vigorous compound strength. Its high figure of merit, drawing nearer 0.8, considered its effective use in a few NASA space missions. Late examinations have revealed the greatest figure of merit upsides of around 1.4 for single stage PbTe-based materials, and 1.8 for homogeneous PbTe-PbSe materials. Silicon-Germanium amalgams (Si1-xGex) are found to be the best TE materials for appliances working under high temperature [23].

6. New Materials

Phonon-glass electron-precious stone (PGEC) materials have a complex intermetallic confine structure, which gives the material great electronic qualities like gem, and simultaneously, a low thermal conductivity, similar to glass. Among other TE materials, half-Heusler composites stand out enough to be noticed with their appealing electrical vehicle properties, somewhat high Seebeck coefficients and rich component mixes [28]. Zintl are ordinarily little bandgap semiconductors with a mind-boggling structure that has the best upsides of figure of merit went from 1 to a pinnacle worth of 1.5. TE oxides, like Ca3Co4O9 (figure of merit almost equal to 1),

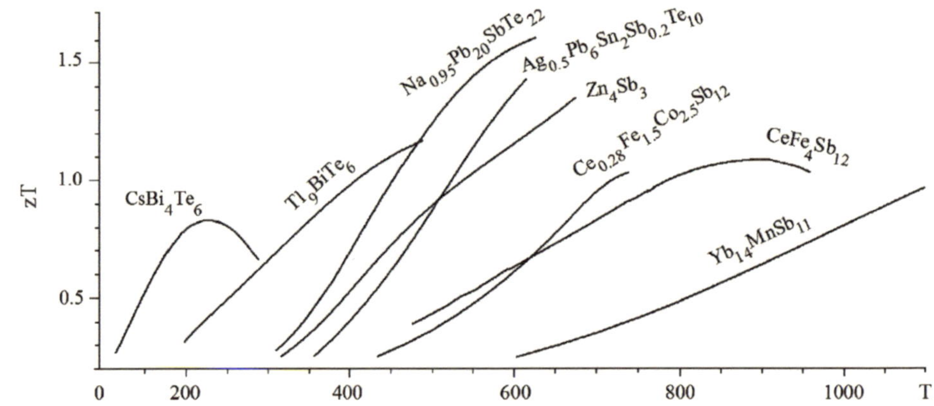

Figure 3.
The p-type TMs of academic importance [12].

are great TE materials, and are environmental and basically stable at high temperatures. The thermoelectric metal chalcogenide has high electrical properties and low conductivity of heat energy, so when exceptional nano-organizing and band designing are utilized, the outcome is a better figure of merit. There are several other TE materials that have been introduced to the field and are of great importance in improving the figure of merit, in result, efficiency [23, 26] (**Figure 3**).

7. Harvesting of body heat

We can use the body heat to harvest thermoelectric energy. Environments that clearly comprise temperature [6] gradients and heat float have the capability to generate electric strength the usage of thermal to electric powered electricity conversion. The temperature distinction gives the capability for green electricity conversion, at the same time as heat float gives the strength. Even with Large heat floats, however, the extractable strength is usually low because of low Carnot and fabric efficiencies [29]. In addition, restrained heat availability may even restrict the strength produced.

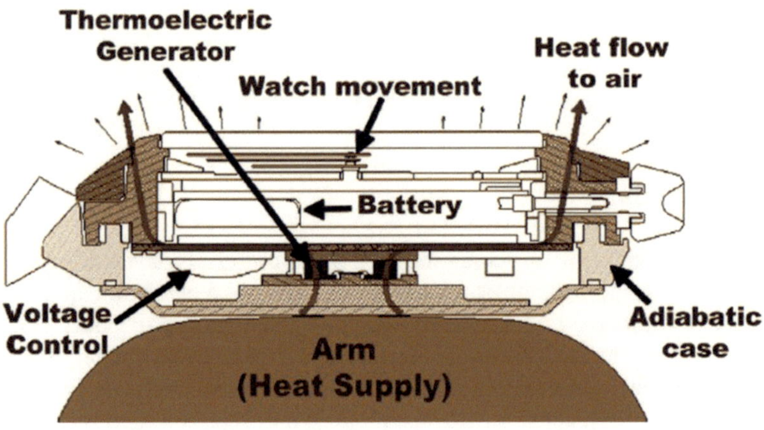

Figure 4.
Schematics of body heat harvesting of Seiko thermic [8, 17].

Nevertheless, for structures with exceptionally low strength necessities, together with faraway wi-fi sensors, thermoelectric electricity harvesting has proven to be a feasible era and promise to emerge as greater typical because the strength necessities for such applications drop. In **Figure 4** a schematic of the watch manufactured by Seiko is shown that operates on thermoelectric energy harvested from body heat [6].

8. Thermoelectric generator

Thermoelectric generators are solid state devices with no moving parts. They are quiet, reliable and scalable, making them ideal for small distributed power generation and energy harvesting. The thermoelectric effect arises because charge carriers in metals and semiconductors move freely like gas molecules and carry charge as well as heat. When a temperature gradient is applied to the material, the charge carriers moving from the hot end will diffuse preferentially towards the cold end. The accumulation of charge carriers results in a net charge (negative for electrons, e, positive for holes, h+) at the cold end, creating an electrostatic potential (potential difference). Thus, an equilibrium is reached between the chemical diffusion potential and the electrostatic repulsion due to the accumulation of charges. This property, known as the Seebeck effect, is the basis of thermoelectric power generation [25] (**Figure 5**).

Thermoelectric device contains multiple thermocouples consisting of n-type (containing free holes) and p-type (containing free holes) thermoelectric elements

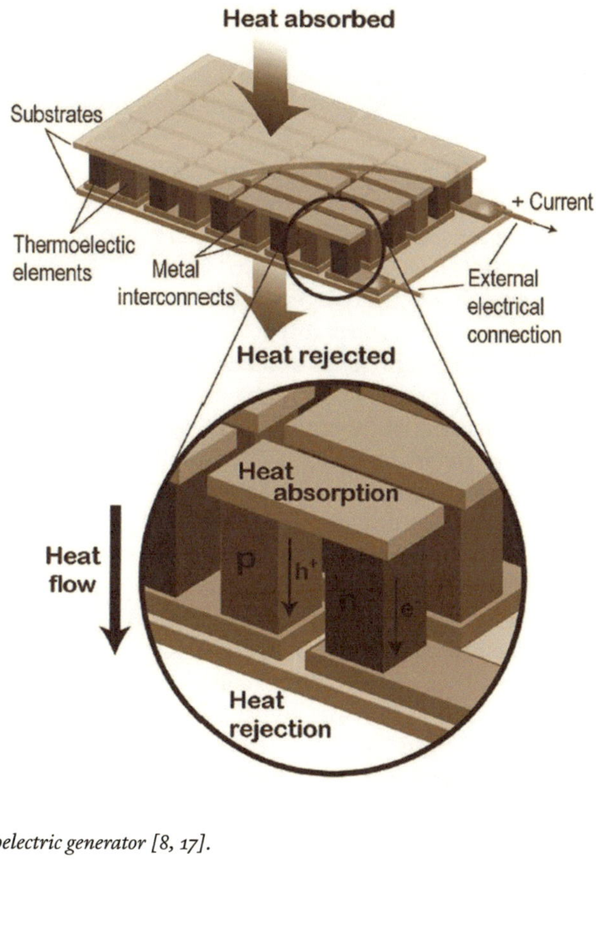

Figure 5.
Schematics of thermoelectric generator [8, 17].

connected in thermoelectric series and in parallel. The best electrolyte is a heavily doped semiconductor. Thermoelectric generators use heat flow through a temperature gradient to power an electrical load through an external circuit. The temperature difference provides the Seebeck effect voltage (Seebeck factor) while the heat flux supports the current, thus determining the output power. The rejected heat must be dissipated by a radiator. The thermoelectric figure of merit of a material depends on the Seebeck coefficient, the absolute temperature, the resistivity and the thermal conductivity of the material. The maximum efficiency of a thermoelectric device is determined by its figure of merit, which is largely the average of the figure of merit of the component materials [26].

9. Applications of TEGs

Thermoelectric applications are ordered by the two impacts portraying the cycle, specifically the Seebeck impact, for any application that creates power with a temperature contrast, and the Peltier impact for any cooling application fueled by an electric flow. The last option case is not analyzed in this article [19, 20]; just applications that produce power are introduced. To create power from a TE module, it is fundamental for there to be a temperature contrast between its hot and cold surfaces. All in all, it is fundamental that the intensity recuperated from the hot source dissipates into the semiconductor components p and n of the module, and afterward to the cold source, which is normally the climate. TEG applications can be arranged into three classifications, contingent upon the idea of the hot source: (i) radioisotope source of heat (ii) regular source of heat (iii) squander source of heat.

A radioisotope thermoelectric generator (RTG) is an atomic electric generator of basic plan. It includes neither a course of combination nor atomic splitting, which would require huge imperatives on the framework, yet the regular rot of a radioactive atom, typically plutonium 238 as plutonium dioxide 238PuO2. As they deteriorate, radioactive atoms discharge heat, some of which is straightforwardly changed over into power [30]. Three regions, specifically the space area, power supply gadgets in distant regions, and the clinical area, have profited from RTGs, albeit the last two regions have not prospered due to the dangers implied in utilizing radioisotopes.

Pouillet, in 1840, utilized the Seebeck impact in making a thermoelectric cell with welded sets of bismuth and copper. The two solderings were drenched in two vessels, one containing dissolving ice and the other high temp water. This mechanical assembly providing a steady wellspring of dynamic power was utilized by the creator to examine the overall laws of flows. These days, thermopiles or [31] TEG thermoelectric generators are intended to supply energy to independent sensors [5], introduced in distant areas subject to serious natural circumstances, i.e., extremely low-temperature and hard to-get to areas, where ordinary environmentally friendly power sources, for example, sunlight based and wind energy, are not routinely accessible. Heat is typically provided by a flameless [32] synergist burner [1–3, 33].

As the intensity of the human body is normal and stable, providing some power in unmistakable applications, for example, clinical ones could be utilized [34, 35]. The human body discharges around 100 W of intensity very still, and 525 W during actual exertion [36]. A few examinations have been led into wearable thermoelectric generators (WTEGs) beginning around 2001 [37], fully intent on subbing lithium-particle batteries [38, 39] as a power hotspot for versatile gadgets. In view of the disadvantages of inflexible modules, i.e., the high thermal resistance among skin and the TEG, adaptable modules are additionally reasonable for power age from body heat, as they can be adjusted to the state of the body, subsequently expanding the valuable [6] surface region for heat catch and diminishing thermal contact resistance.

A sun powered thermoelectric generator (STEG) is a framework intended to recuperate heat from sun oriented radiation and convert it into power utilizing a thermoelectric generator (TEG) [5]. It is turning into an innovative other option, and is rivaling the predominant sun powered photovoltaic frameworks regardless of its low transformation effectiveness contrasted with photovoltaic innovation [40]. STEGs are characterized by the sort of optical sensors utilized, specifically, an optical fixation framework or not. Optical focus sensors are normally barrel shaped focal points, Fresnel focal points, illustrative mirrors, level mirrors or allegorical concentrators. Non-concentrated arrangements are somewhat restricted to level plate authorities cleared or not emptied, and vacuum tubes.

A gigantic measure of poor quality waste intensity is delivered into the climate, with next to no endeavor at heat recuperation. Throughout the course of recent years, much effort has been made to work on the proficiency of thermoelectric innovation utilized in heat recuperation applications. It is worked that TE innovation could be effectively adjusted to the actual boundaries, like the temperature, strain, and intensity move liquid, of a given intensity recuperation application. Squander heat recuperation utilizing thermoelectric innovation can be partitioned into two principal gatherings, as follows—(i) squander heat recuperation from industry and homes (ii) squander heat recuperation from transport frameworks.

10. Thermoelectric modules

In order to expand the use of thermoelectricity, it is essential to manufacture standard thermoelectric modules of different sizes, accessible to all [8]. A typical thermoelectric generator (TEG) [5] module consists of 10–100 n-type and p-type thermoelectric elements, electrically connected in series and in parallel, alternating between two ceramic layers. The p-n pairs are connected by conductive poles that are connected to the elements through a low-melting solder (PbSn or BiSn). When a temperature gradient occurs between its two junctions, the TEG converts thermal energy into electrical energy according to the principle of the Seebeck effect. This flat bulk architecture is most used and marketed. However, in some applications, flat shapes are impractical. Indeed, it is difficult to adapt the heat source to the thermoelectric device, which makes it more expensive, heavier and bulkier. Therefore, other designs are being explored to overcome these disadvantages. Although most are cylindrical in shape, they are still not marketed. This is limited research, unlike thick and thin films and flexible TE devices, which are developed more efficiently. Two ceramic plates act as support for the module and insulator, but Thermistor reduces the efficiency of the module. Since then, a number of studies have suggested the concept of direct contact thermoelectric generators (DCTEGs), which are characterized by one of the module surfaces in direct contact with the heat source and another surface in direct contact with coolant flow. There are several manufacturing technologies for TE modules such as plate lithography, raised process, flash evaporation, thin film evaporation, photolithography and etching, screen printing, sputtering, distribution foil printing, engineering, etc. spark plasma sintering, direct current (DC) magnetic sputtering, and printing processes. A key challenge during the development of TEGs was the degradation of the original properties caused by thermal fatigue, which is caused by thermal expansion and thermal shock, respectively. This degradation can be sudden or gradual, resulting in reduced life and performance. In fact, during normal operation of TE devices, the shields are periodically heated and cooled and undergo thermal expansion. The TE material connected to these screens can experience different expansion effects from temperature sources, causing increased stress at the interface between them. These stresses are often the main cause of mechanism failure, and are therefore the main reason why TE materials are not sintered and incorporated into the shunts.

11. Conclusions

The extraordinary idea of utilizing thermoelectric generators, to supply an electric flow with a temperature contrast of any little worth and over a wide temperature range, has made them the center answer for specific energy issues in regards to drive age and intensity recuperation in a static and non dirtying way, considerably under outrageous natural circumstances. The low effectiveness of this transformation innovation has restricted its turn of events, besides in specific areas where the benefits of TEGs are better over different advances. The utilization of thermoelectricity in different lab and modern areas has brought about there being alternate points of view. It has made critical progress in certain applications and absolute disappointment in others. The flow examinations concerning thermoelectric generators are centered around the improvement of new productive thermoelectric materials to conquer the downsides of the interconnected electrical and warm properties of these materials, and new plans of thermoelectric generators that permit better reconciliation into energy change frameworks, according to the perspective of productivity and the ecological effect. Interest in this innovation has been resuscitated with the presence of nanotechnology, which has made it conceivable to cross the authentic boundary of figure of merit<1, bringing about a remarkable expansion in distributions in this field. In this audit, the best in class of thermoelectric generators, applications and late advancement are completely revealed. Crucial information on the thermoelectric impact, essential regulations, and boundaries influencing the productivity of traditional and new thermoelectric materials, are completely examined. The uses of thermoelectricity are assembled into three fundamental areas. The primary gathering manages the utilization of intensity transmitted from a radioisotope to supply power to different gadgets, with just space ending up the area in which thermoelectricity was effective. In the subsequent gathering, a characteristic intensity source can be helpful for delivering power, however this gathering of thermoelectricity is currently at a lacking stage due to low transformation productivity, leaving applications still at research facility level. The third gathering is advancing at a high velocity, chiefly on the grounds that the examinations are financed by states and additionally vehicle producers whose last point is to diminish vehicle fuel utilization, and thusly relieve ozone depleting substance outflows.

Author details

Hiranmoy Samanta*, Kamal Golui and Soumyadeep Mukherjee
Gargi Memorial Institute of Technology, Shree Chaitanya College, Habra, Kolkata, India

*Address all correspondence to: hiru.samanta@gmail.com

References

[1] Elsheikh MH, Shnawah DA, Sabri MFM, Said SM, Hassan MH, Bashir MBA, et al. A review on thermoelectric renewable energy: Principle parameters that affect their performance. Renewable and Sustainable Energy Reviews. 2014;**30**:337-355

[2] Arutyunov VS, Lisichkin GV. Energy resources of the 21st century: Problems and forecasts. Can renewable energy sources replace fossil fuels? Russ. Chemical Reviews. 2017;**86**:777-804

[3] Oral AY, Oral ZBB. 3rd International Congress on Energy Eciency and Energy Related Materials (ENEFM2015): Proceedings. Oludeniz, Turkey; 2015

[4] Schierning G, Chavez R, Schmechel R, Balke B, Rogl G, Rogl P. Concepts for medium-high to high temperature thermoelectric heat-to-electricity conversion: A review of selected materials and basic considerations of module design. Translational Material Research. 2015;**2**:025001

[5] Thermoelectric Generators (TEGs). Gentherm Global Power Technologies. 2018. Available online: http://www.genthermglobalpower.com/products/thermoelectric-generators-tegs [Accessed: June 22, 2018]

[6] Nozariasbmarz A, et al. High power density body heat energy harvesting. ACS Applied Materials & Interfaces. 2019;**11**(43):40107-40113

[7] Narducci D, Giulio F. Recent advances on thermoelectric silicon for low-temperature applications. Materials. 2022;**15**:1214

[8] Zoui MA, Bentouba S, Stocholm JG, Bourouis M. A review on thermoelectric generators: Progress and applications. Energies. 2020;**13**:3606

[9] Nozariasbmarz A, Collins H, Dsouza K, Polash MH, Hosseini M, Hyland M, et al. Review of wearable thermoelectric energy harvesting: From body temperature to electronic systems. Applied Energy. 2020;**258**:114069

[10] Levinsky P. Synthesis, Characterization and Optimization of New Thermoelectric Materials. Prague: Université de Lorraine; 2018

[11] Mahan GD. Introduction to thermoelectrics. APL Materials. 2016;**4**:104806

[12] Polozine A, Sirotinskaya S, Schaeffer L. History of development of thermoelectric materials for electric power generation and criteria of their quality. Materials Research. 2014;**17**(5):1260-1267

[13] Leonov V. Thermoelectric energy harvesting of human body heat for wearable sensors. IEEE Sensors Journal. 2013;**13**:5. DOI: 10.1109/JSEN.2013.2252526

[14] Cronin B. Vining: An inconvenient truth about thermoelectrics. Nature Materials. 2009;**8**:83-85

[15] Martin P. Thermoelectric materials and applications. News. Bulletin. 2005;**2005**:30

[16] Riffat SB, Ma X. Thermoelectrics: A review of present and potential applications. Applied Thermal Engineering. 2003;**23**:913-935

[17] Jeffrey Snyder G. Thermoelectric Energy Harvesting

[18] Kasap S. Thermoelectric Effects in Metals: Thermocouples. Canada: Department of Electrical Engineering University of Saskatchewan; 2001

[19] Altenkirch WWE, Gehlho GR. Thermo-electric heating and cooling body. U.S. Patent 1120781A. 1914;**15**

[20] BJohn F, Harry B, Reggie S. Thermoelectric cooling units. U.S. Patent 2932953A. 1960

[21] Radousky HB, Liang H. Energy harvesting: An integrated view of materials, devices and applications. Nanotechnology. 2012;**23**:502001

[22] Goldsmid HJ, Douglas RW. The use of semiconductors in thermoelectric refrigeration. British Journal of Applied Physics. 1954;**5**:386-390

[23] Selvan KV, Hasan N, Ali MSM. State-of-the-art reviews and analyses of emerging research findings and achievements of thermoelectric materials over the past years. Journal of Electronic Materials. 2018;**48**:745-777

[24] Ioffe AF. Power Application Principles of Semiconductor Thermopiles. USSR Academy Science, 1950. Available online: https://ia800607.us.archive.org/35/items/A.F.IoffeSemiconductorThermoelementsAndThermoelectricCoolingInfosearch1957/A.F.%20Ioffe-Semiconductor%20Thermoelements%20and%20Thermoelectric%20Cooling%20%20-Infosearch%20(1957)_text.pdf

[25] Ma T, Qu Z, Yu X, Lu X, Wang Q. A review on thermoelectric-hydraulic performance and heat transfer enhancement technologies of thermoelectric power generator system. Thermal Science. 2018;**22**:1885-1903

[26] Horie S. Thermoelectric Energy Conversion Systems. Tokyo, Japan: REALIZE Science & Engineering; 1995. pp. 112-115

[27] Bin Masood K, Kumar P, Singh RA, Singh J. Odyssey of thermoelectric materials: Foundation of the complex structure. Journal of Physical Communication. 2018;**2**:062001

[28] Xu Q, Qu S, Ming C, Qiu P, Yao Q, Zhu C, et al. Conformal organic–inorganic semiconductor composites for flexible thermoelectrics. Energy & Environmental Science. 2020;**13**:511-518

[29] Sun T, Zhou B, Zheng Q, Wang L, Jiang W, Snyder GJ. Stretchable fabric generates electric power from woven thermoelectric fibers. Nature Communications. 2020;**11**:572

[30] Yang J, Caillat T. Thermoelectric materials for space and automotive power generation. MRS Bulletin. 2006;**31**:224-229

[31] Thomson W. On the dynamical theory of heat transfer. Transactions of Royal Society Edinburg. 1851;**3**:91-98

[32] BioLite-ROW, BioLite Outdoor & O-Grid Energy|Rest-Of-World, BioLite-ROW. Available online: https://row.bioliteenergy.com/ [Accessed: June 25, 2018].

[33] Bonin R, Boero D, Chiaberge M, Tonoli A. Design and characterization of small thermoelectric generators for environmental monitoring devices. Energy Conversion and Management. 2013;**73**:340-349

[34] Francioso L, De Pascali C, Farella I, Martucci C, Creti P, Siciliano PA, et al. Flexible thermoelectric generator for ambient assisted living wearable biometric sensors. Journal of Power Sources. 2011;**196**:3239-3243

[35] Leonov V, Vullers R. Wearable electronics self-powered by using human

body heat: The state of the art and the perspective. J. Renew. Sustain. Energy. 2009;**1**:62701

[36] Bhatnagar V, Owende P. Energy harvesting for assistive and mobile applications. Energy Science Engineering. 2015;**3**:153-173

[37] Siddique ARM, Mahmud S, Van Heyst B. A review of the state of the science on wearable thermoelectric power generators (TEGs) and their existing challenges. Renewable and Sustainable Energy Reviews. 2017;**73**:730-744

[38] Karthikeyan V, Surjadi JU, Wong JC, Kannan V, Lam K-H, Chen X, et al. Wearable and flexible thin film thermoelectric module for multi-scale energy harvesting. Journal of Power Sources. 2020;**455**:227983

[39] Han S. Wearable thermoelectric devices. In: Thermoelectric Thin Films. Cham, Switzerland: Springer Science and Business Media LLC; 2019. pp. 31-42

[40] Stoll BL, SmithiD A, Deinert M. Potential for rooftop photovoltaics in Tokyo to replace nuclear capacity. Environmental Research Letters. 2013;**8**:014042